Ion Exchange and Solvent Extraction

Ion Exchange and Solvent Extraction

A Series of Advances

Volume 14

edited by

Arup K. SenGupta
Lehigh University
Bethlehem, Pennsylvania

Yizhak Marcus
The Hebrew University of Jerusalem
Jerusalem, Israel

Jacob A. Marinsky
Founding Editor

CRC Press
Taylor & Francis Group
Boca Raton   London   New York

CRC Press is an imprint of the
Taylor & Francis Group, an **informa** business

CRC Press
Taylor & Francis Group
6000 Broken Sound Parkway NW, Suite 300
Boca Raton, FL 33487-2742

First issued in paperback 2019

© 2001 by Taylor & Francis Group, LLC
CRC Press is an imprint of Taylor & Francis Group, an Informa business

No claim to original U.S. Government works

ISBN-13: 978-0-8247-0508-4 (hbk)
ISBN-13: 978-0-367-39744-9 (pbk)

**Visit the Taylor & Francis Web site at
http://www.taylorandfrancis.com**

**and the CRC Press Web site at
http://www.crcpress.com**

Preface

I read somewhere that *chemists don't fade away; they just seek new and exciting equilibrium.* Professor Jacob A. Marinsky, the founding editor of the Ion Exchange and Solvent Extraction series, has decided to step down from the helm of editorship. Needless to say, Jack is to embrace a more meaningful equilibrium outside the realm of ion exchange. I am stepping into his shoes and am fully aware that those shoes are too big to fill. On behalf of the entire "ion exchange" community, I would like to thank and applaud Jack for his leadership, farsightedness, and effort in shouldering the responsibility of editing this series for over 30 years through 13 volumes, which have immensely benefited two generations of professionals. The first volume was entitled *Ion Exchange* and Professor Yizhak Marcus was a chapter contributor. Later, Marcus joined as the co-editor and the name of the series changed to Ion Exchange and Solvent Extraction. I look forward to working with Professor Marcus and maintaining the high quality and visibility of the series.

Over the past three decades, theory and application of ion exchange have greatly advanced and touched almost every industrial sector—from mining to microelectronics. Nevertheless, new ground is still being broken in this field, in terms of both syntheses of novel materials and application opportunities in new areas. This fourteenth volume of the Ion Exchange and Solvent Extraction series encompasses state-of-the-art development in an area that may rightly be called *environmental separation by ion ex-*

change. The volume brings together polymer chemists, technologists, scientists, and engineers to provide comprehensive coverage of the emerging role of ion exchange in environmental processes.

Traditionally, each polymeric ion exchanger contains one chemically distinct repeating functional group. In Chapter 1, Alexandratos presents different classes of bifunctional polymers and elaborates how bifunctionality can enhance sorption affinity toward target solutes and at the same time improve accessibility and uptake kinetics.

Ion exchange resins have long been known to remove dissolved ions from solutions. Ion exchangers may also be used to recover valuable species from precipitates or sparingly soluble ores that dissociate only slightly to give traces of valuable ions in solution. In Chapter 2, van Deventer et al. provide a detailed account of prevailing equilibrium relationships and supporting experimental data for systems in which leaching and ion exchange take place simultaneously.

Selective separation of valuable metals and their possible recovery is an area of considerable interest. Chelating exchangers in the form of spherical beads are used in packed-bed columns for such separation processes. In Chapter 3, Ritchie and Bhattacharyya discuss at length the potential and relative advantages of using microfiltration or nanofiltration membrane processes for metals removal after covalently attaching polyamino carboxylate groups to the membrane substrate.

Increased use of biorenewable materials in engineered processes is a desirable goal. In this regard biosorbents that are chemically processed dead microorganisms or seaweeds have shown very high metal selectivity. Cases of uranium cation and gold cyanide anion biosorption and modeling of engineered systems are discussed by Volesky et al. in Chapter 4.

In polymeric chelating exchangers, chelating ligands are covalently bonded to polymeric support or matrix. Preformed silica gels or organoceramic polymers may also serve as the substrate for covalent attachment of metal-selective ligands or deposition of solvent extractant. In Chapter 5, Tavlarides and Lee present the preparation methodologies and metals separation properties of functionalized organo-ceramic sorbents that are compatible with packed-bed processes.

To date, use of ligand exchange processes is mostly limited to analytical separations. In Chapter 6, SenGupta presents a special class of polymeric ligand exchanger (PLE) with high affinities toward many environmentally significant anionic ligands.

In separation science, molecular recognition has recently drawn much attention. One of the promising ways to create a highly selective sorbent is through the "molecular imprinting technique." In Chapter 7, Goto presents sorption behaviors of Zn(II)-imprinted polymers synthesized through (1) post-irradiation of the polymer with gamma rays and (2) the design of novel host molecules.

Inorganic iron oxides are selective toward metal ions primarily at near-neutral to slightly alkaline conditions. Also, their physical configurations (powdered forms) render them unusable in packed-bed columns. In Chapter 8, Kney and SenGupta present methodologies of preparing hybrid iron-rich all-inorganic granular particles that are metal-selective at pH as low as 4.0 and amenable to efficient regeneration.

Arup K. SenGupta
Yizhak Marcus

Contributors to Volume 14

Spiro D. Alexandratos Department of Chemistry, University of Tennessee, Knoxville, Tennessee

Dibakar Bhattacharyya Department of Chemical and Materials Engineering, University of Kentucky, Lexington, Kentucky

P. G. R. de Villiers Department of Chemical Engineering, University of Stellenbosch, Stellenbosch, South Africa

Masahiro Goto Department of Chemical Systems and Engineering, Kyushu University, Fukuoka, Japan

Arthur D. Kney Department of Civil and Environmental Engineering, Lafayette College, Easton, Pennsylvania

J. S. Lee Department of Chemical Engineering and Materials Science, Syracuse University, Syracuse, New York

L. Lorenzen Department of Chemical Engineering, University of Stellenbosch, Stellenbosch, South Africa

Hui Niu Department of Chemical Engineering, McGill University, Montreal, Canada

Stephen M. C. Ritchie Department of Chemical and Materials Engineering, University of Kentucky, Lexington, Kentucky

Arup K. SenGupta Department of Civil and Environmental Engineering, Lehigh University, Bethlehem, Pennsylvania

Lawrence L. Tavlarides Department of Chemical Engineering and Materials Science, Syracuse University, Syracuse, New York

Jannie S. J. van Deventer Department of Chemical Engineering, University of Melbourne, Victoria, Australia

Bohumil Volesky Department of Chemical Engineering, McGill University, Montreal, Canada

Jinbai Yang Department of Chemical Engineering, McGill University, Montreal, Canada

Contents

Contents of Other Volumes

Ion Exchange and Solvent Extraction

1

Polymer-Supported Reagents
The Role of Bifunctionality in the Design of Ion-Selective Complexants

Spiro D. Alexandratos

University of Tennessee, Knoxville, Tennessee

The importance of multifunctionality in the preparation of ion-selective polymers is evident from the structure of enzymes where specific metal ions are bound through cooperative interactions among different amino acids. In synthetic polymers, ionic selectivity is enhanced when a chemical reaction is superimposed on an ion exchange process. The concept of *reactive ion exchange* has been extended through the synthesis of crosslinked polymers whose metal ion selectivity is a function of reduction, coordination or precipitation reactions as determined by various covalently bound ligands. Development of three classes of *dual mechanism bifunctional polymers*, a new series of bifunctional diphosphonate polymers, and novel bifunctional ion-selective polymers with enhanced ionic accessibility will be presented. Bifunctional anion-selective polymers, interpenetrating polymer networks, and high-stability solvent-impregnated resins will also be examined.

I. INTRODUCTION

The cooperativity of amino acid ligands in binding metal ions has been the subject of extensive research and is an important factor in their selectivity. For example, the zinc ion in carbonic anhydrase II (CAII) is bound by three separate histidine residues in the enzyme [1]. This has led to CAII being used in fluorescence-based biosensors for the detection of zinc ions in solutions [2] with concentrations as low as the picomolar range [3]. The microenvironment around the ligands is an additional factor in the ionic selectivity of the enzyme; amino acids with aromatic moieties near the binding site enhance the selectivity for zinc ions [4] by influencing the rigidity of the groups available for binding [5]. It is this multifunctionality of enzymes that allows for selective metal ion interactions and that becomes the central theme in our research aimed at designing highly selective polymer-supported reagents.

II. ION EXCHANGE RESINS AND THE DEVELOPMENT OF ION-SELECTIVE POLYMERS

The widely available polystyrene-based sulfonic acid resin has very similar affinities for a wide range of cations [6] due to similar free energy of reactions [7]. This ion exchange resin can be used to remove metal ions such as copper from solutions, but only when the level of competing ions such as calcium and iron are low [8]. The need for ion-selective polymer-supported complexants has thus long been recognized.

A. Reactive Ion Exchange

Helfferich proposed that selectivity is possible when ion exchange occurs along with a chemical reaction involving the metal ion [9]. Eleven examples of ion exchange coupled with chemical reactions were subdivided into four categories: in type I processes, counterions from the ion exchanger reacted with co-ions in solution (this included neutralization reactions); in type II processes, solution counterions reacted with immobilized exchange sites on the polymer; in type III processes, undissociated exchange sites were ionized by reaction with solution co-ions (this would include salt formation reactions); and in type IV processes, undissociated exchange sites were changed from one undissociated form to another by reaction with the solution co-ions. A reaction thus occurs which results in the exchanging ions no longer retaining their original chemical properties. This was later defined as *reactive ion exchange*, which involves the exchange of ions between phases coupled with reactions that yield new products or otherwise alter initially present species allowing for complete removal of the target ion from its originating phase [10]. Three methods were proposed in which ions could be chemically changed: (1) the charge on the ion is increased or decreased while retaining its original sign; (2) the charge on the ion is decreased or increased to the opposite sign; and (3) the charge on the ion is eliminated to give a zero-valent species. These concepts were applied and extended in the development of *reactive polymers*, to be detailed in Section III.

B. Soluble Phosphorus-Based Metal Ion Complexants

An understanding of the development of polymer-supported reagents within the context of reactive ion exchange requires an overview of the soluble organophilic complexants used in solvent extraction processes. Solvent extraction is a popular technique due to its versatility, rapid rates of complexation, high selectivity, and the extensive library of complexants that has been synthesized [11]. Organophilic metal ion complexants used in solvent extraction processes have been prepared with a wide array of functional groups. Particularly versatile are those complexants with phosphorus-based ligands, including phosphoric acids, phosphonic acids, phosphate esters, and phosphine oxides. They have been studied over many years [12] and are still in current use.

Organophosphorus acids have significant affinities for rare earth and transition metal ions. Di(2-ethylhexyl)phosphoric acid (DEHPA) is an important complexant used under different conditions. A solution in hexane is able to remove trace levels of neodymium(III) from acidic solutions of varying pH where the ionic strength is kept constant at 0.1 M with sodium nitrate [13]. A kerosene solution complexes yttrium(III) from highly acidic solutions [14] and iron(III) from aqueous chloride solutions [15]. A kerosene solution of DEHPA that has been partially converted to its sodium salt with base becomes a more effective complexant for americium(III) from nitric acid solutions [16]. A comparative study of partially neutralized phosphoric, phosphonic, and phosphinic acids dissolved in kerosene showed that the type of phosphorus acid was important to determining the affinity toward cobalt and nickel ions from acidic solutions, with the organophilic phosphinic acid being most effective [17].

Neutral organophosphorus complexants are another class of compounds that have been extensively studied. Tri-n-octyl phosphine oxide (TOPO) is one such compound. It has, for example, high affinities for both beryllium(II) and aluminum(III) and can be used to separate the former from the latter by adjusting the pH of the aqueous phase [18]. Dihexyl-N,N-diethylcarbamoylmethylphosphonate and octylphenyl-N,N-diisobutylcarbamoylmethylphosphonate are two compounds with the general structure $R'_2P(O)CH_2C(O)NR_2$ that are very effective actinide complexants from highly acidic solutions [19]. From less acidic solutions, 2-(dihexylphosphino)-pyridine N,P-dioxide is a promising complexant that is more selective for europium(III) than ytterbium(III) or cerium(III) [20].

It is the ability of phosphorus to form both ion exchange and neutral coordinating ligands with affinities for different metal ions that depend on their specific structural characteristics which became most important in our development of ion-selective polymer-supported reagents (*vide infra*).

C. Synergistic Interaction Between Ligands

Under certain conditions, solvent extraction with an organic solution of two different complexants yields a much greater level of extraction than would be expected from the behavior of each complexant alone. This phenomenon was termed *synergistic extraction.* An early observation involved the extraction of uranyl ions from acidic solutions when a complexant capable of ion exchange such as DEHPA was combined with a neutral coordinating complexant such as TOPO or tributyl phosphate (TBP) [21]. Synergism has been identified in the extraction of lithium when an aqueous solution is contacted with DEHPA and TBP in kerosene [22]; when the solution containing low (ppm) levels of lithium, sodium, potassium, magnesium, and calcium ions is contacted with an organic phase containing DEHPA alone, the affinity series is Ca(II) > Mg(II) > Li(I) > K(I) > Na(I) with 42.6% of the lithium being complexed; when the organic phase consists of both DEHPA and TBP, the amount of lithium complexed increases to 76.1%, while complexation of the other metal ions is unchanged. The same combination of DEHPA and TBP is synergistic in the extraction of iron(III) from aqueous chloride solutions [23]. Diaryldithiophosphinic acids and TOPO are synergistic in their complexation of americium(III) and europium(III) from nitric acid solutions as concentrated as 1 M, whereas the phosphinic acids have no affinity for either ion when used alone [24].

Synergism is also evident when the neutral coordinating compound is a crown ether rather than an organophosphate. There is a synergistic interaction when dialkylphosphoric acids are combined with stereoisomers of dicyclohexano-18-crown-6 in the complexation of strontium and barium ions, but not calcium ions [25]. In another example, the combination of dibenzo-18-crown-6 with 8-hydroxyquinoline in chloroform is synergistic for the extraction of cobalt(II) from aqueous solutions [26].

The concept of synergistic combinations between ligands with acid sites and neutral coordinating ligands to augment the extent of metal ion complexation, when combined with the concept of reactive ion exchange, led to the design and development of polymer-supported reagents with cooperative ligand interactions for enhanced metal ion affinities and selectivities. These cooperative interactions were tailored to produce a chemical or coordinative reaction.

III. BIFUNCTIONAL ION-SELECTIVE POLYMER-SUPPORTED REAGENTS

A. Dual Mechanism Bifunctional Polymers

Bifunctional polymers were synthesized with two different ligands immobilized onto a polymer support in order to determine whether synergistic interactions between the ligands could lead to enhanced metal ion affinities and selectivities. These polymers were collectively categorized as *dual mechanism bifunctional polymers* (DMBPs). The underlying theme is that one ligand acts through a relatively aspecific mechanism to allow metal ions access into the polymer matrix, followed by a selective recognition mechanism acting on the targeted metal ion [27]. The DMBPs are divided into three classes: ion exchange is the access mechanism in each class because the ligand's hydrophilicity makes it compatible with the metal ion's hydration shell; the recognition mechanism then defines the selective reaction of each class.

1. Class I DMBPs

The first class of DMBPs consists of a reduction reaction as the recognition mechanism combined with ion exchange for rapid access into the polymer matrix. The Class I DMBPs focus on the synthesis of primary and secondary phosphinic acid ligands on polystyrene (Fig. 1). The resin is an effective ion exchanger for transition metal ions, as seen in studies with Zn(II) in the presence of a large excess of sodium ions [28,29]. The redox mechanism becomes operative when the resin is contacted with Hg(II) or Ag(I) ions. In these reactions, the metal ion enters the matrix by ion exchange, followed by its reduction to the zero-valent metallic state once it approaches the P–H bond (which is the source of electrons in the reduction reaction). The primary phosphinic acid reduces Hg(II) with a 1:1 stoichiometric ratio (Fig. 2) and Ag(I) with a 1:2 ratio (Fig. 3). Reduction of Hg(II) is more rapid than Ag(I) since only one Hg(II) ion must be proximate to the P–H bond for reduction to occur. The primary phosphinic acid ligand is then oxidized to phosphonic acid [30]. The resin is stable in 4 M nitric acid at ambient temperature. The minimum metal ion reduction potential that still allows for reduction by phosphinic acid is approximately 0.3 V [31].

In the absence of cations capable of undergoing reduction [Ag(I), Hg(I), Hg(II), and Cu(II)], the phosphinic acid resin was found to ion exchange in the order Pb(II) > Mn(II) ≫ Cd(II) > Ca(II) > Zn(II) >

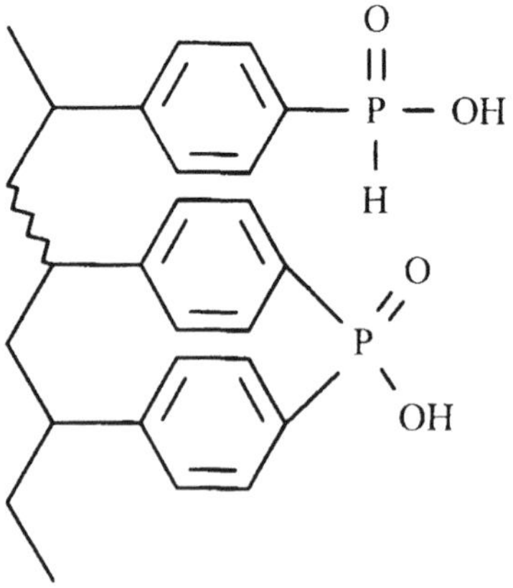

Figure 1   Class I DMBP with primary and secondary phosphinic acid ligands.

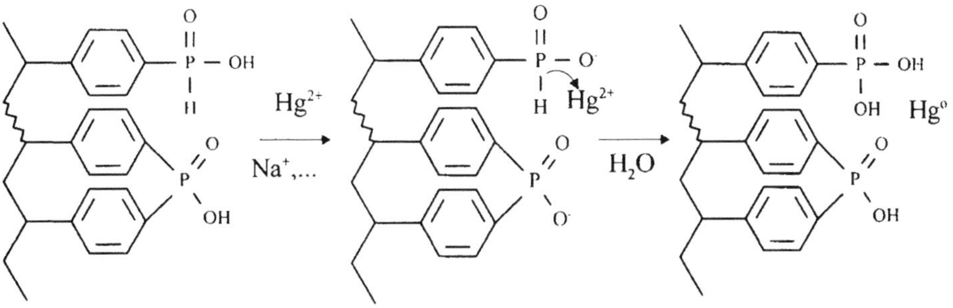

Figure 2   Reduction of Hg(II) by the primary phosphinic acid ligand in aqueous solution.

Mg(II) > Ni(II) > Ba(II) > Sr(II) > Co(II) ≫ Fe(III) ≫ Cs(I) ≫ Li(I) > Na(I) > Rb(I) [32]. Its ability to complex lanthanides and actinides was compared to the sulfonic acid resin and quantified by the distribution coefficient, D, defined as mmol M^{n+} per $g_{dry\ resin}$/mmol M^{n+} per mL_{soln} [33]. A correlation of log D versus pH for uptake of europium and americium from nitric acid solutions shows that when the acidity is varied from 4 M to 0.25 M, the slopes of the lines for both ions with the phosphinic acid resin are 1.75, while the slopes for the sulfonic acid resin are 3.0. This indicates that the sulfonic acid exchanges with the trivalent ions, while the phosphinic acid exchanges with the metal nitrate complexes (Fig. 4). When sodium nitrate is added in an amount to give a constant 4 M nitrate background in all solutions of varying pH, the uptake for

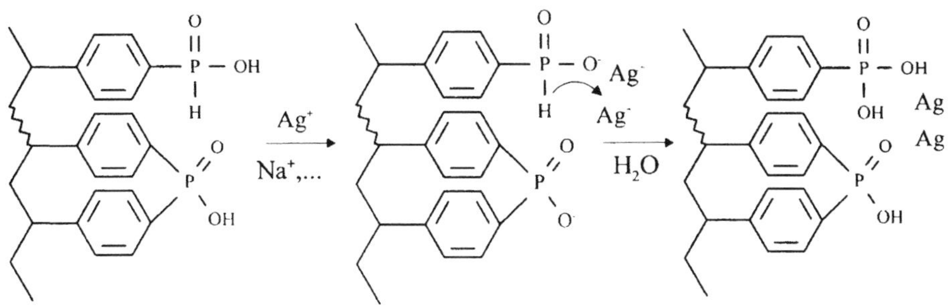

Figure 3   Reduction of Ag(I) by the primary phosphinic acid ligand in aqueous solution.

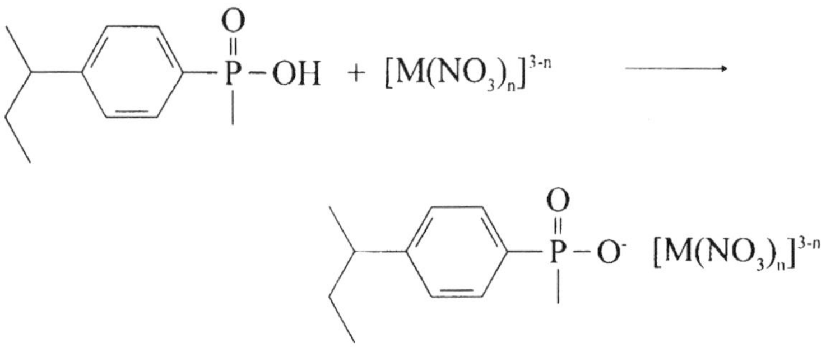

Figure 4 Complexation of metal nitrate by the phosphinic acid resin.

the phosphinic acid resin remains almost unchanged, as do the slopes of the log D versus pH correlation [1.90 and 1.97 for Eu(III) and Am(III), respectively], while the uptake for the sulfonic acid resin falls to negligible levels and the slopes are near zero for both metal ions. This underscores the greater selectivity of the phosphinic acid resin over a wide range of solution conditions since the sulfonic acid resin exchanges for the sodium ions present in large excess, which decreases the distribution coefficients for the targeted metal ions. The phosphinic acid resin was also superior in its uptake of both uranium and thorium at high acid concentrations and of plutonium at all acid concentrations in the presence of sodium ions.

In order to better understand metal ion uptake kinetics by the phosphinic acid resin, Ni(II) was studied in solutions of constant pH at 2 and 60°C [34]. Equilibrium was reached in 1 h at the lower temperature and in less than 15 min at the higher temperature. Ion exchange was the only mechanism operative as determined by reanalysis of the acid capacity after uptake and regeneration (a redox reaction gives an acid capacity higher than the original value). The amount of Ag(I) removed from aqueous solutions also increased as a function of temperature; the resin acid capacity after regeneration confirmed activity of the redox reaction.

2. Class II DMBPs

The second class of DMBPs couples ion exchange as the access mechanism with coordination as the recognition mechanism, the objective being to determine whether synergism could be observed when ligands were immobilized on a polymer support. Examples of Class II DMBPs are those with phosphonate monoethyl/diethyl ester ligands, phosphonic acid/tertiary amine ligands, and carboxylic acid/pseudocrown ether ligands (Fig. 5) [35,36]. In studies with the monoethyl/diethyl ester resin and americium nitrate, it was found that metal ion uptake in high levels of sodium nitrate is dependent on the percentage of diester, monoester, and diacid ligands on the phosphoryl groups [35]: the monofunctional phosphonate mono-ethyl ester resin gave the highest Am(III) uptake of all resins studied, followed by resins with varying amounts of monoethyl/diethyl ester li-

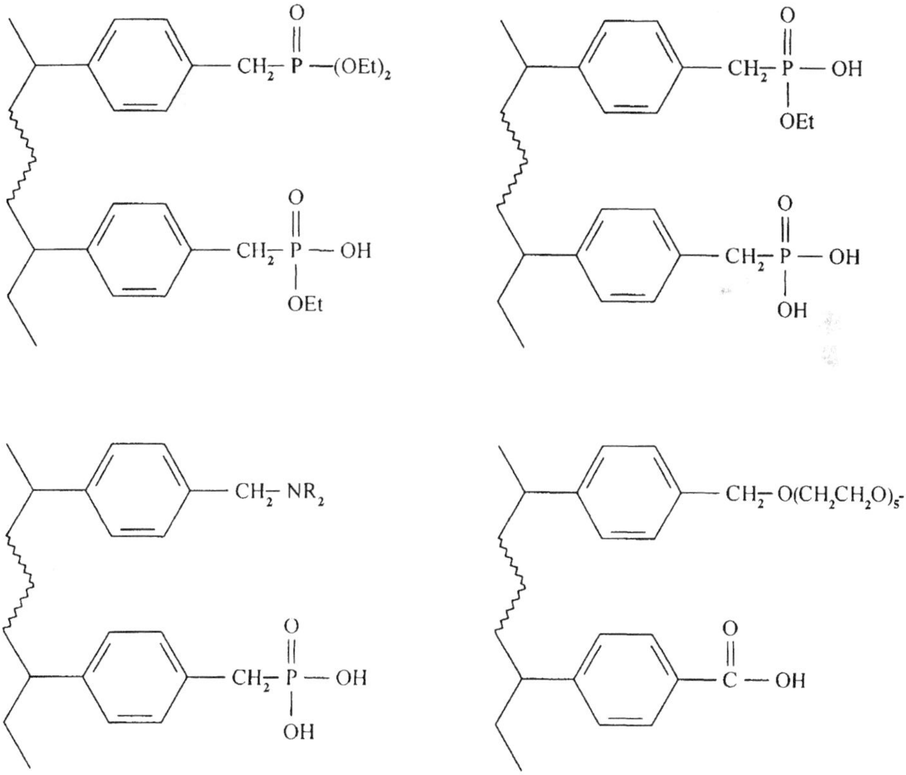

Figure 5 Class II DMBPs with phosphonate monoester/diester, phosphonic acid/phosphonate monoester, phosphonic acid/tertiary amine, and carboxylic acid/pseudocrown ether ligands.

gands, while resins with monoester/diacid ligands gave lower results that were similar to those of the monofunctional phosphonic acid resin. The phosphonate diethyl ester resin had almost no affinity for the americium ions. These results are in contrast to those found with silver ions, which show clear evidence of synergism with the bifunctional resin (see Section III.A.4 on comparative study).

3. Class III DMBPs

The third class of DMBPs couples a precipitation reaction as the recognition mechanism with ion exchange for enhanced metal ion access (Fig. 6) [37]. A phosphonic acid ligand allows for ion exchange and a quaternary amine ligand leads to precipitation via its associated anion once the targeted metal ion exchanges into the resin. This results in both specificity and removal of the metal ion from solution by intraresin precipitation, which isolates the precipitated salt within the bead (Fig. 7). Different cation selectivities are found depending on whether the associated anion is sulfate, thiocyanate, or iodate. The metal can be recovered by solubilizing the salt with the appropriate eluents.

4. Comparative Study of DMBP–Metal Ion Affinities

A study of the ionic affinities displayed by a series of resins was completed in order to gain a better understanding of the mechanisms by which

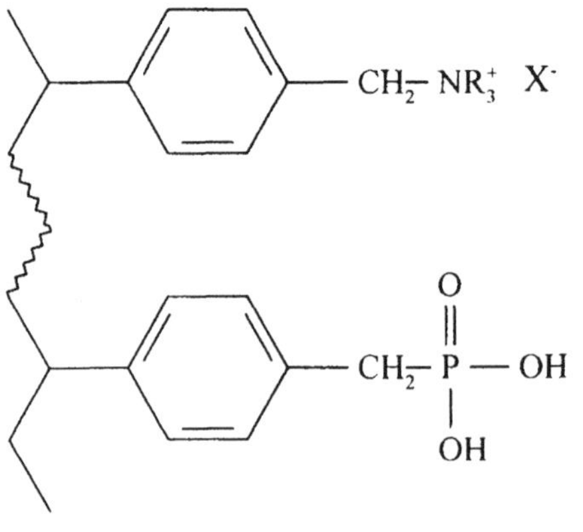

Figure 6   Class III DMBP with phosphonic acid/quaternary amine ligands.

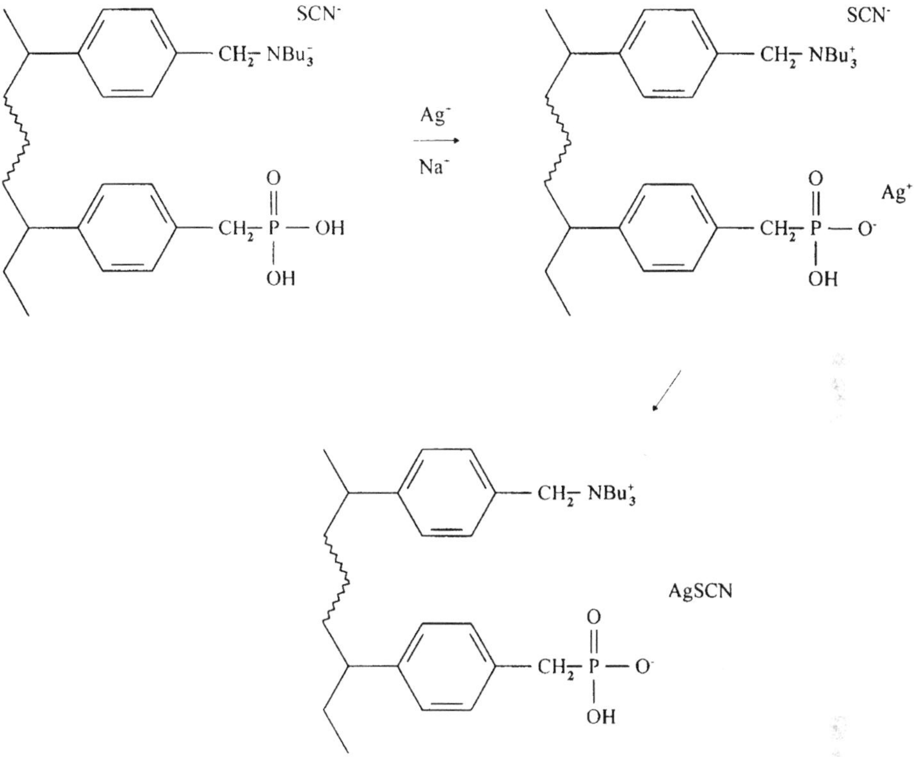

Figure 7 Ion exchange/precipitation reaction of the class III DMBP with silver ions in aqueous solution.

DMBPs interact with metal ions [38]. Ten resins (sulfonic acid, phosphinic acid, phosphonic acid, dimethylamine, phosphonic acid/dimethylamine, phosphonate monoethyl ester, dimethyl ester, diethyl ester, dibutyl ester, and monoethyl/diethyl ester) were contacted with 10^{-4} N solutions of five different metals [Fe(III), Hg(II), Ag(I), Mn(II), and Zn(II)], each in four different concentrations of nitric acid (0.2, 1.0, 2.0, and 4.0 M) with the ionic strength maintained at a constant 4 M through the addition of sodium nitrate.

The sulfonic acid resin complexed low levels of all the transition metal ions because its ion exchange sites were saturated with the sodium ions present in much higher concentrations or inoperative (in the 4 M nitric acid solution) because the acid strength was far greater than the sites' pKa. On the other hand, the phosphinic acid resin was far more selective and displayed a series of Fe > Hg ≫ Ag > Mn > Zn in all solutions. Its affinity for Fe(III) and Hg(II) is significantly higher than for other metals. While the high affinity for mercury was expected due to intervention by the reduction reaction, the affinity for Fe(III) was less expected and later understood to be due to coordination through the phosphoryl oxygen. This ability to convert from an ion exchange to a coordination mechanism for certain ions, especially in solutions of high acidity, is an important property of resins with phosphoryl ligands. The phosphonic acid resin has the same general trend as the phosphinic acid resin except for a large decrease in uptake of mercuric ions due to the absence of the reduction mechanism; the high affinity for Fe(III) is retained due, again, to coordination by the phosphoryl oxygen.

The dimethylamine resin complexed moderate levels of only Hg(II) and Ag(I), while the bifunctional acid/amine resin gave results similar to those of the phosphonic acid resin. In this case, the ligands on the bifunctional resin behave independently, and the distribution coefficients are seen to be reliable indicators of the extent to which ligands cooperate in the binding of metal ions. This conclusion became particularly important in results with the ester resins.

The phosphonate monoethyl ester resin performs like the phosphonic acid resin except for its higher affinity for Ag(I). The three dialkyl ester resins all retain a high affinity for Ag(I), especially from the more acidic solutions, and show an affinity series of Ag ≫ Hg > Fe ≫ Zn ~ Mn. Most importantly, the bifunctional monoethyl/diethyl ester resin had the highest affinity for Ag(I) of all the resins studied. Its level of complexation was much greater than could be expected from results with the monofunctional analogs (e.g., from the 4 M HNO_3 solution, the distribution coefficients for the monoethyl/diethyl, monoethyl, and diethyl resins were 2980, 500, and 550, respectively). The results thus clearly indicate participation of a *supported ligand synergistic interaction*. The basis for this mechanism is a compatibility between the metal ion and the ligands based on the principles of hard–soft acid–base theory with superimposed steric constraints from the coordinating ligand.

B. A General Method of Enhancing Ionic Accessibility in Crosslinked Polymers

The metal ion affinities of monofunctional resins with coordinating ligands such as the phosphonate diester resins indicate that while such resins can be selective in their ionic interactions, consistent with the behavior of soluble analogs, the kinetics of complexation can be slow due to diffusional limitations into the polymer matrix. A general method for enhancing the kinetics of complexation was thus developed that could be applied to a wide range of coordinating ligands immobilized on crosslinked polymer supports. This was accomplished by utilizing the sulfonic acid ligand as the principal access ligand, based on the observation that the sulfonic acid resin is highly hydrophilic and has rapid ion exchange kinetics under different conditions.

Monofunctional phosphonic acid resins crosslinked with 2 and 12% divinylbenzene (DVB) were found to quantitatively complex 0.0001 N Eu(III) in 0.01 and 0.10 M nitric acid, but this decreased significantly as the acidity of the background solution increased to 0.50 and 1.0 M nitric acid [39]. The reason typically given for this behavior was that the complexation mechanism changes from ion exchange to coordination at the phosphoryl oxygen as the solution pH falls below the pKa of the acid ligand, and this gives a lower metal ion affinity due to competition by protons in solution for P=O sites. However, an alternative explanation would be that the polymer does not remain hydrated in low pH (high ionic strength) solutions, and this prevents metal ion access due to collapse of the matrix. When the resin was sulfonated to give a bifunctional sulfonic/phosphonic acid resin (Fig. 8), complexation of Eu(III) increased to 95–100% for all solutions. The inherently hydrophilic sulfonic acid ligand thus hydrates the resin, and this prevents the matrix from collapsing

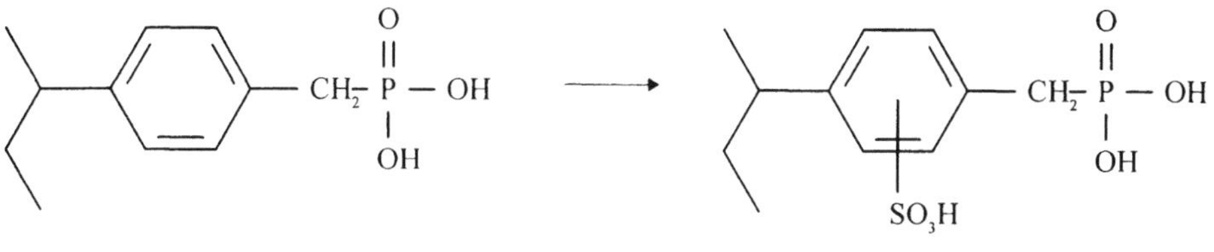

Figure 8 Sulfonation of polystyrene-supported phosphonic acid.

in highly acidic solutions. The same results were found when 0.40 M sodium nitrate was added to each of the four acid solutions, indicating that the high levels of Eu(III) complexation in the bifunctional resin are not due to exchange with the sulfonic acid ligand since it will have exchanged with the sodium ions present in large excess. The sulfonic acid ligand thus allows for access of all ions into the matrix, while the phosphonic acid ligand complexes the target ion through a coordinative mechanism when solution conditions are unfavorable for ion exchange.

The effect of bifunctionality is especially pronounced as matrix crosslinking increases and contact with the ion-containing solution decreases [40]. When phosphonic acid resins with crosslinking at 5, 12, and 20% DVB are contacted with a dilute solution of europium nitrate in 1 M nitric acid for 0.5 h, the amount complexed decreases from 31 to 0.61%. On the other hand, the bifunctional sulfonic/phosphonic acid resins complex 94–95% Eu(III) under the same conditions. Bifunctionality thus enhances accessibility even into highly crosslinked resins and does so to a much greater extent than by making the monofunctional resins macroporous [40].

Sulfonation may be accomplished at low temperatures with chlorosulfonic acid after the ion-selective ligand has been immobilized. It does not require the presence of unsubstituted phenyl rings since the electrophilic substitution reaction occurs on rings that have been phosphorylated [39]. Additionally, it is not necessary to sulfonate all of the aromatic rings in order to increase the complexation rate. Sulfonating 20% of the aromatic rings in a fully functionalized phosphinic acid resin increased the distribution coefficient of Eu(III) in a 1 M nitric acid solution from 61 to 220 at a 0.5-h contact time [41].

IV. A NEW FAMILY OF SELECTIVE POLYMERS BASED ON THE GEM-DIPHOSPHONATE LIGAND

Hydroxyethane-1,1-diphosphonic acid is known to be an effective complexant for a variety of metal ions, even from acidic solutions [42]. As a result, it was decided to convert it to vinylidene-1,1-diphosphonic acid and prepare its polymerized analog. While the monomer was too sterically hindered to homopolymerize with a free radical initiator, formation of the tetraester allowed for free radical copolymerization with acrylonitrile, styrene, and DVB. After hydrolysis of the ester to the diphosphonic acid, it was found that the polymer had slow rates of complexation. However, immobilizing sulfonic acid groups on the phenyl rings provided by incorporating styrene gave a bifunctional polymer whose ionic selectivity is coupled with rapid complexation rates [43]. This polymer has been commercialized as Diphonix® ion exchange resin (Fig. 9).

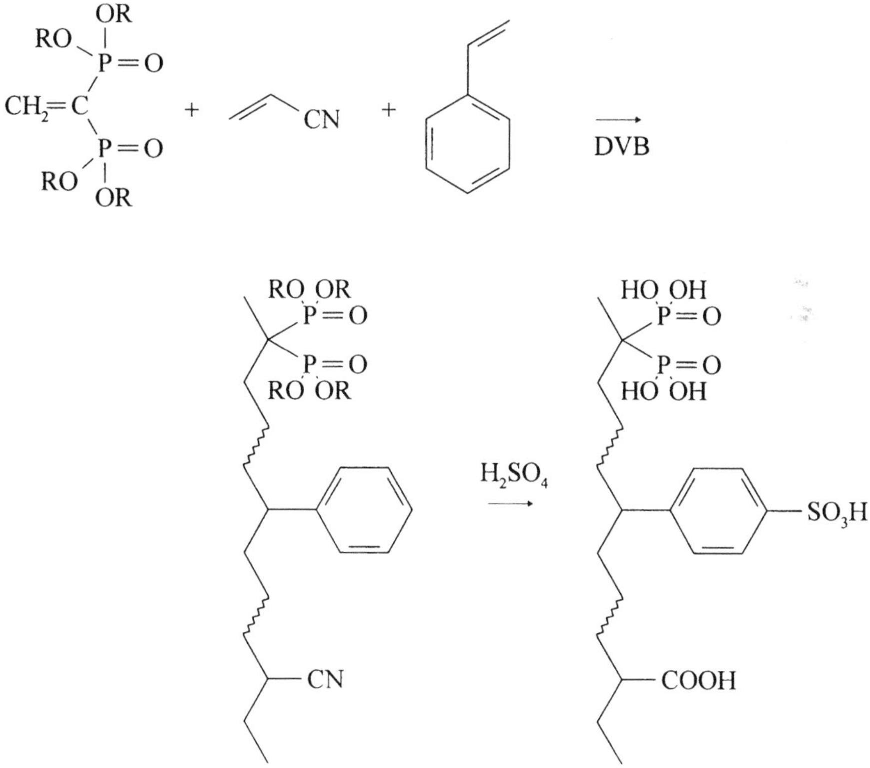

Figure 9 Synthesis of Diphonix.

A. Diphonix

Diphonix forms highly stable metal complexes with tetra- and hexavalent actinides in acidic solutions by chelating the actinides with the gem-diphosphoryl group [44]. The distribution coefficients for U(VI) and the tetravalent actinides [Th(IV), Np(IV), and Pu(IV)] fall within the remarkably high range of 10^4 to 10^7 in solutions with nitric acid concentrations up to 10 M [45]. The slopes of the log D versus pH correlation are all significantly less than expected from the ionic charge, thus indicating a dominant coordinative component to the interaction. Only Am(III) uptake shows a slope indicating exchange with the trivalent ion. Comparing the kinetics of Am(III) uptake between Diphonix and its unsulfonated analog confirms the importance of the sulfonate ligand to the rapid rates of complexation [45]: Diphonix reaches 99.9% of its equilibrium uptake value at a 10-min contact time, while the unsulfonated analog requires days to reach equilibrium.

A comparative study of the log D/pH correlations with Am(III), U(VI), and Pu(IV) ions involving Diphonix, its monophosphonic acid analog, and the sulfonic acid resin highlights the unique ability of Diphonix to complex very high levels of metal ions due to the geminal arrangement of the diphosphoryl groups. Diphonix significantly outperforms the other resins in the complexation of uranium and plutonium ions; for example, in 10 M nitric acid, the Pu(IV) distribution coefficients for Diphonix, the monophosphonic acid resin, and the sulfonic acid resin are 17,000, 1000, and 20, respectively. Interestingly, Diphonix and the sulfonic acid resin behave identically toward the trivalent americium ion in solutions less acidic than 4 M HNO_3, though as the acid concentration increases from 4 M, the sulfonic acid resin continues to complex less Am(III), while Diphonix stays constant at the level found at 4 M acid. The monophosphonic acid resin complexes the least amount of Am(III) at all acid concentrations, indicating that it is both a far less effective coordinator than Diphonix and less effective at ion exchange [46]. The presence of high concentrations of sodium and calcium ions does not affect the uptake by Diphonix of tri-, tetra-, and hexavalent actinides, which can allow for decontamination of numerous types of solutions [47].

A study of the uptake of transition metal ions shows that Diphonix strongly coordinates Fe(III) over the range of 0.02 to 10 M HNO_3 with a maximum at approximately 0.1 to 5 M HNO_3 [48]. Over the same range of acidity, Ca(II), Co(II), and Zn(II) are complexed by Diphonix with a slope of 2 in the log D/pH correlation, indicating that ion exchange is the sole sorption mechanism. From solutions at a pH between 5 and 8, Diphonix has a much higher affinity than the sulfonic acid resin for Cr(III), Mn(II), Co(II), Ni(II), Cu(II), Zn(II), Cd(II), and Pb(II) and similar affinities for Al(III), Hg(II), Sn(II), and Sb(II). Rate experiments with Co(II) in 0.05 M HNO_3, Zn(II) in 0.10 M HNO_3, Fe(III) in 1 M HNO_3, and Co(II) in pH 6 aqueous solution show that Diphonix has very fast rates of complexation for transition metal ions [45].

Diphonix has a high affinity for indium from solutions of 1 M H_2SO_4, even in the presence of 0.1 M Na_2SO_4 [49]. Additionally, approximately 96% of the indium is complexed from 1 M H_2SO_4 solutions containing 1 mM In(III) and 1 mM Zn(II) or Cu(II). The resin displayed rapid rates of complexation under both batch and continuous flow conditions.

B. Diphonix II

An alternate preparation of polymers with immobilized gem-diphosphonate ligands was developed by functionalizing poly(vinylbenzyl chloride) with tetraisopropyl methylenediphosphonate (Fig. 10) [50]. A gel copolymer crosslinked with 2% DVB gives 55% substitution with the diphosphonate ligand, while a crosslinked vinylbenzyl chloride (VBC)–styrene copolymer (2:1 molar ratio) gives 72% functionalization, indicating that the bulky tetraisopropyl ligands sterically hinder complete reaction. When the crosslink level of the VBC–styrene copolymer increases to 5% DVB, functionalization decreases to 25%; this value is sensitive to the surface area of the copolymer since the macroporous analog reaches 45% func-

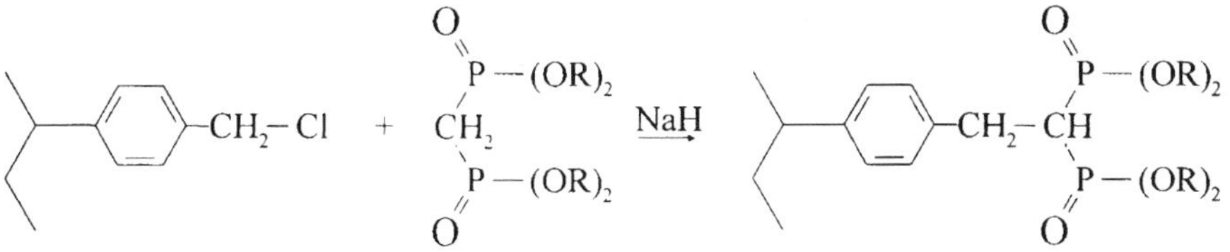

Figure 10 Synthesis of Diphonix II.

tionalization. Macroporosity influences metal ion uptake by the mono-functional diphosphonate resin: at a 0.5 h contact time, the 5% DVB gel resin complexes 11.4% Eu(III) from a 1 M HNO_3 solution, while the macroporous resin complexes 69.2%. However, converting the gel to a bifunctional sulfonic/diphosphonic acid resin increases the Eu(III) uptake from 11.4 to 100%, again pointing to the importance of adding an access ligand to the resin.

C. Diphonix-A

A modified Diphonix was successfully synthesized that had an affinity for both cations and anions, which can be important to the treatment of nuclear wastes that contain both transuranium ions and ^{99}Tc (present as the pertechnetate anion) [51]. This analog, labeled Diphonix-A, was pre-pared by replacing the sulfonic acid ligand with either a quaternary trialkyl-ammonium group (Diphonix-A, Type 1) (Fig. 11) or an alkylated pyri-dinium group (Diphonix-A, Type 2) (Fig. 12). Both Type 1 and Type 2 resins have a high affinity for TcO_4^-, reaching equilibrium within five

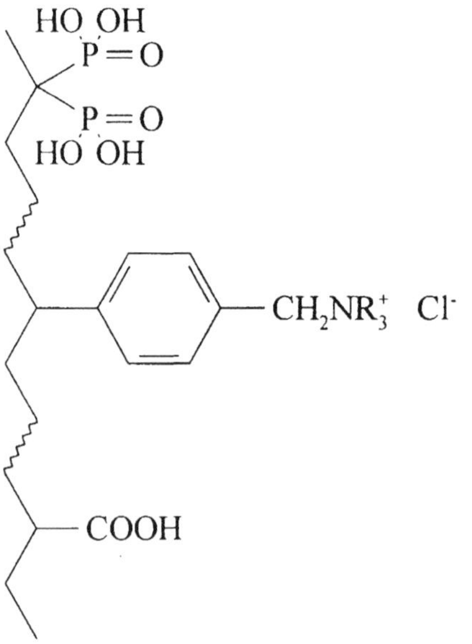

Figure 11   Diphonix-A, Type 1 resin.

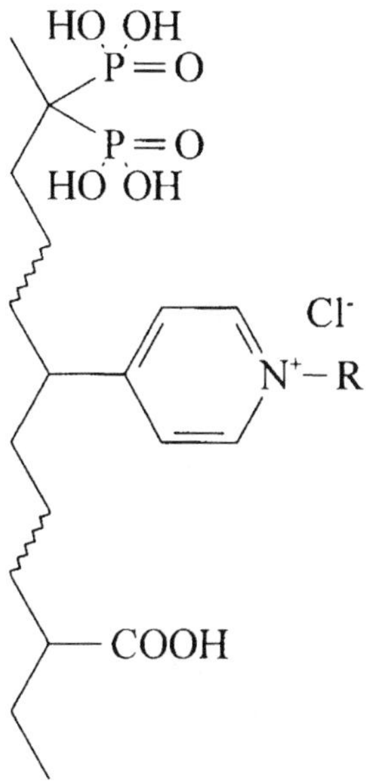

Figure 12   Diphonix-A, Type 2 resin.

minutes. The Type 1 resins, however, have an apparently decreased affinity for actinide ions, which is actually due to slow complexation kinetics, while the Type 2 resins retain the rapid equilibration rate expected from the behavior of Diphonix and an apparent high affinity for actinide ions. It is concluded that the trialkylammonium ligand interacts electrostatically with the diphosphonate ligand and this interaction needs to be broken before complexation with the actinide ions can occur; the rapid equilibration rates with the Type 2 resins indicate that this interaction is absent between the pyridinium and diphosphonate ligands. Diphonix-A Type 2 resins may thus have an important role in radionuclide separations given an affinity for TcO_4^- from acidic solutions that is comparable to that of commercial anion exchange resins and an affinity for actinide ions that is comparable to Diphonix.

D. Diphosil

An inorganic analog of Diphonix has been prepared by immobilizing the diphosphonate ligands on silica gel [52]. Vinylbenzyl chloride was grafted on to silica gel after modifying the latter with a toluene solution of trichloroethylsilane and trichlorovinylsilane; reaction with tetraisopropyl methylenediphosphonate and subsequent hydrolysis gave Diphonix on silica, or Diphosil (Fig. 13). It may be treated as solid waste after use in actinide separations since only 10% of its total weight is organic, which results in minimal generation of gaseous by-products due to radiolytic decomposition.

While Diphosil has a lower phosphorus capacity than Diphonix (0.68 versus 1.64 mmol/g, respectively, for two typical batches), the former has its capacity entirely on the surface layer provided by the porous silica support, while the latter has its capacity distributed throughout the gel polymer bead. Diphosil thus has a higher surface capacity of ion-binding ligands on the surface than does Diphonix, which obviates the need

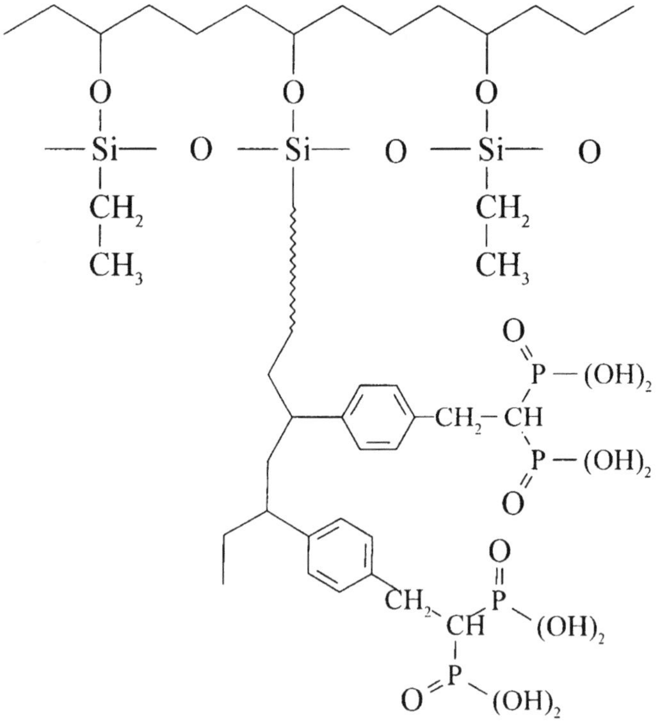

Figure 13   Diphosil.

for Diphosil to have a sulfonic acid ligand for enhanced access. This is supported by rate experiments with solutions of Am(III), U(VI), Th(IV), and Fe(III) that show that both polymers reach 99% of the equilibrium value of the distribution coefficient within 10 min of contact. Both polymers have comparable affinities for Am(III) and U(VI), while Diphosil shows a much higher affinity for Pu(IV); for example, the Pu(IV) distribution coefficients in 1 M HNO$_3$ are 10^5 and 10^6 for Diphonix and Diphosil, respectively. The affinity for Th(IV) is very high for both polymers with the distribution coefficient being 10^6 for both in 3 M HNO$_3$; results with Diphosil are less dependent on solution acidity with the effect that its distribution coefficients are greater than Diphonix in solutions more acidic than 3 M and lower than Diphonix in solutions less acidic than 3 M.

E. Diphonix-CS

The alkaline liquid portion in high-level nuclear waste storage tanks contains high levels of ^{137}Cs and ^{90}Sr. While Diphonix has a high affinity for Sr(II) from alkaline solutions, it has a relatively low affinity for Cs(I). Combining both cesium and strontium selectivity on a single resin would facilitate treatment of the alkaline supernate for waste minimization. In another example of how coupling the diphosphonate ligand with a second ion-selective ligand extends the conditions under which Diphonix may be applied, polyphenolic groups were immobilized alongside the diphosphonate ligand (Fig. 14) [53]. Tetraisopropyl vinylidene diphosphonate was first copolymerized with vinylbenzyl chloride, the —CH$_2$Cl groups converted to aldehyde moieties, and the polyphenolic ligands grafted there by reaction with phenol and formaldehyde. The resin prepared in this manner, referred to as Diphonix-CS, did not require sulfonation since it is to be used under alkaline conditions wherein the diphosphonate ligands are ionized, thus swelling the resin and enhancing ionic accessibility. Its affinity for both cesium and strontium is very high: from 1 M NaOH, the distribution coefficients are 3000 and 15,000 for Cs(I) and Sr(II), respectively, while the corresponding values for Diphonix are 10 and 25,000, respectively. Results are consistent with the two ligands in Diphonix-CS acting independently in their complexation of Cs(I) and Sr(II) ions.

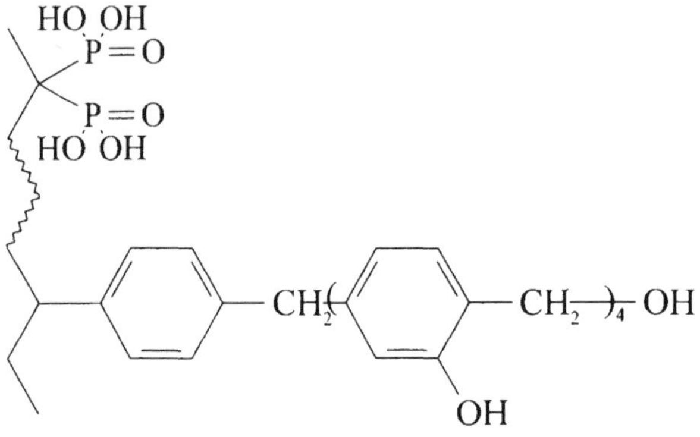

Figure 14   Diphonix-CS.

Diphonix is thus found to be a versatile resin wherein the conditions under which it can be utilized may be extended by modifying either the support or ligand structure.

V.   INTRALIGAND COOPERATION: POLYMERS WITH IMMOBILIZED KETOPHOSPHONATE LIGANDS

The metal ion affinities displayed by Diphonix are much higher than those found with other immobilized ligands and may be attributed to intraligand cooperation wherein the ligating sites are connected through a common atom, allowing for chelate formation with the substrate. This is in contrast to the Class II DMBPs where interligand cooperation is responsible for the supported ligand synergistic interaction described earlier. Interligand cooperation effects a significant increase in metal ion affinities relative to results in its absence, but intraligand cooperation may permit still greater affinities.

In order to more fully explore the general concept of intraligand cooperation, a series of polymers with immobilized ketophosphonate ligands were synthesized. The carbonyl and phosphoryl moieties were positioned adjacent to each other in the α-ketophosphonate and then separated by one and two methylene moieties to give the β- and γ-keto-

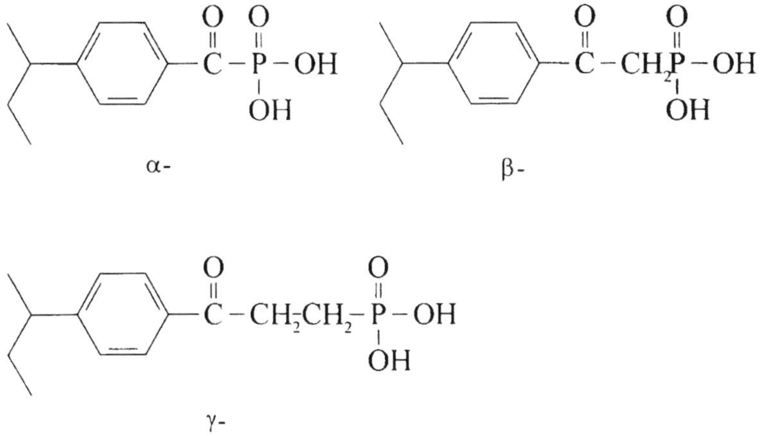

Figure 15 Ketophosphonate resins used in the study of intraligand cooperation.

phosphonates, respectively (Fig. 15) [54]. The metal ion affinities were
also compared to the monophosphonate ligand, for which intraligand cooperation is not possible.

Metal ion affinities with the four polymers for Pb(II), Cu(II), Cd(II),
Co(II), Ag(I), and Eu(III) were quantified from solutions of 0.01, 0.10,
and 1.0 M nitric acid. In each case where a trend is evident, the α-ketophosphonate outperforms the β-ketophosphonate, which, in turn, is far
more effective than the γ-ketophosphonate; the γ-ligand is comparable to
the monophosphonate. The results with Pb(II) and Cu(II) are representative: from 0.10 M nitric acid, the Pb(II) distribution coefficients for the
α-, β-, γ-, and monophosphonate ligands are 1060, 366, 40.8, and 59.8,
respectively (97.5%, 90.7%, 53.1%, and 55.6% complexed), while for
Cu(II) the values are 224, 64, 15.1, and 18.3 (89.5%, 63.0%, 29.6%,
and 27.8% complexed).

The results are consistent with the absence of a significant interaction between the carbonyl and phosphoryl moieties in the γ-ketophosphonate, given that it behaves similarly to the monophosphonate ligand, due
to an unacceptably large loss in entropy upon formation of the seven-
membered ring that would result if the groups cooperated in binding the
metal ion. However, interaction between the carbonyl and phosphoryl

moieties clearly occurs in the α- and β-ketophosphonates to form five- and six-membered rings, respectively, with the former being somewhat favored because of the smaller loss in rotational entropy. Interligand cooperation is thus an important concept for application to selective metal ion complexation studies. The ketophosphonates and diphosphonates offer a range of ionic affinities for application to metal ion separations under different solution conditions and are an important set of polymer-supported reagents for continued study.

VI. BIFUNCTIONAL ANION-SELECTIVE POLYMER-SUPPORTED REAGENTS

The design of polymer-supported reagents that are selective for metal oxyanions in the presence of common anions such as chloride, nitrate, and sulfate presents a unique challenge. An important application of a suitable resin would be removal of the radioactive pertechnetate anion from groundwater with which it is contaminated in certain areas at a concentration that is four to five orders of magnitude less than that of other anions naturally found in groundwater [55]. Commercially available anion exchange resins, including Purolite A-520-E, Ionac SR-6, Reillex HPQ, Dowex 1-X8, and Amberlite IRA-904, are not sufficiently selective for this application [55].

Metal oxyanions such as TcO_4^- are characterized by a single negative charge that is delocalized over a number of atomic centers resulting in a species that is highly polarizable with a low charge-to-volume ratio. It is reasonable to expect, based on the principles of hard–soft acid–base theory [56], that a polymer-supported reagent with polarizable ligands of the opposite charge would be selective for the metal oxyanion. Since quaternary ammonium ion ligands can be prepared with a low charge-to-volume ratio by increasing the size of the groups on the nitrogen, trihexylamine was immobilized on poly(vinylbenzyl chloride) beads to give the trihexylammonium ion ligand, which has a single positive charge associated with a very large ligand. Contact studies showed the resin to be selective for TcO_4^-, but with a low exchange rate due to the ligand's hydrophobicity [55]. While the trimethyl or triethylammonium resins reach equilibrium within 24 h (distribution coefficients of 1690 and 5690, respectively), the resin with trihexylammonium ligands does not reach equilibrium even after a contact time of 336 h, though the distribution coefficient of 28,000 at that point indicates a very high affinity for the pertechnetate ion. Given the selectivity of the trihexylammonium ion ligand toward TcO_4^-, it was evident that coupling that ligand with an access ligand could allow for rapid and selective complexation of oxyanions and thus give the first anionic analog of the dual mechanism bifunctional polymers. Studies undertaken to combine the selectivity of the longer alkyl ammonium ligands with hydrophilic (and aspecific) ammonium ligands having less bulky alkyl groups found that the bifunctional anion exchange resin that best combined TcO_4^- selectivity with rapid rates of complexation was formed by successive reactions of macroporous beads with trihexylamine and then with triethylamine (Fig. 16). This resin had a TcO_4^- distribution coefficient at 24 h of 15,810, compared to 1570 for the pure trihexylamine resin. Field tests under continuous conditions at the site of contaminated groundwater showed the bifunctional resin to perform approximately five

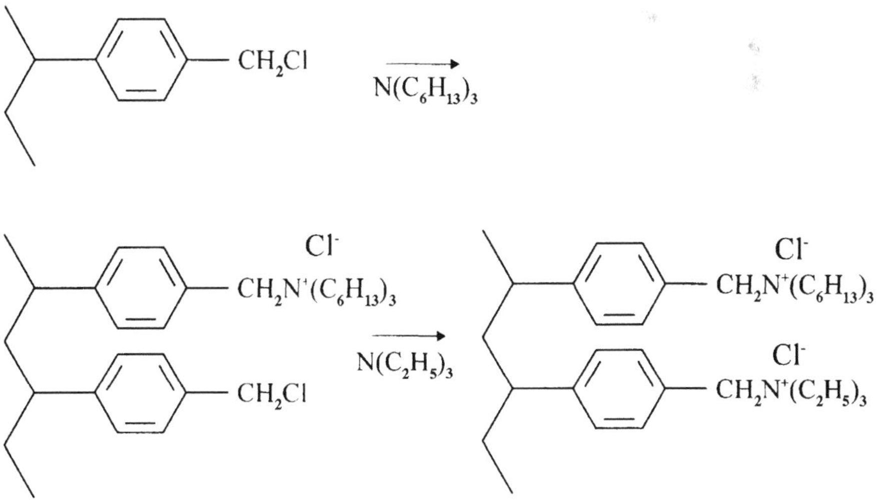

Figure 16 Synthesis of an anion-selective dual mechanism bifunctional polymer.

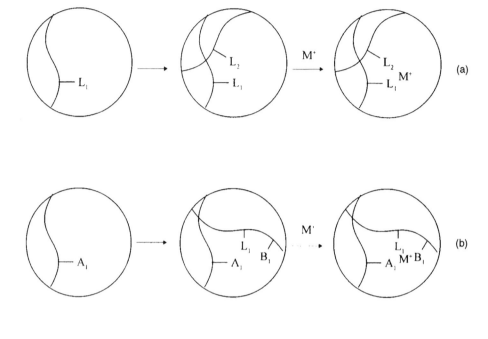

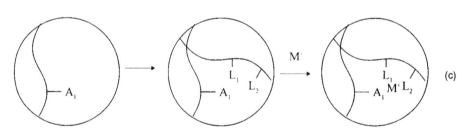

Figure 17   Schematic representation of interpenetrating polymer networks.

times better than one of the best commercially available resins, Purolite A-520-E: at 1% breakthrough of pertechnetate, the bifunctional resin had treated 580,000 bed volumes, while the Purolite resin had treated only 105,000 bed volumes [57]. This behavior was confirmed with other monovalent oxyanions and established the generality of the concept of access/

recognition in dual mechanism bifunctional polymers by extending it to anions.

VII. BIFUNCTIONAL INTERPENETRATING POLYMER NETWORKS

Interpenetrating polymer networks (IPNs) are formed when the properties of one polymer are modified by the addition of a second polymer [58]. In one synthetic method, an initially formed crosslinked polymer sorbs a monomer that is subsequently polymerized to form a second network that is entangled within it but not bonded to it. Including a crosslinking agent with the second monomer crosslinks the second network and prevents any phase separation that could occur between the two networks.

Interpenetrating polymer networks offer a novel means of preparing bifunctional polymers and probing for supported ligand synergistic interactions. An ion-complexing ligand, L_1, can be immobilized on an initially formed network and a second ligand, L_2, can be immobilized on the second network (Fig. 17a). The resulting affinities can be contrasted with results from the isolated L_1 and L_2 polymers. If the first network consists of nonbinding groups A_1 and the second network is formed with ion-binding ligands L_1, then the effect of the microenvironment formed by network A_1 on the metal ion affinity of ligand L_1 can be quantified. The *microenvironmental effect* can be varied with different nonbinding groups $(A_2, A_3, \ldots)$ as well as by forming the second network with different ratios of L_1 and another nonbinding group B_1 (Fig. 17b). The nonbinding groups (A_n, B_n) are able to modify the ion-binding affinities of ligand L_1 by varying the polymer's polarity or hydrophilicity. If the second network formed within the nonbinding network A_1 consists of two binding ligands, L_1 and L_2, then the microenvironmental effect can be superimposed on any supported ligand synergistic interaction (Fig. 17c).

In studying IPNs where the microenvironment of ligand L_1 was varied by introducing differing amounts of nonbinding group B_1 within an initial A_1 network, N-vinylimidazole (VI) and ethyl acrylate (EA) were polymerized within an initial network of crosslinked polystyrene [59]. Five IPNs were prepared with final VI/EA ratios of 100:0, 77:23, 53:

47, 27:73, and 0:100. Imidazole was thus the ion-binding ligand, and its microenvironment within the nonpolar polystyrene was modified with the carboethoxy group—a moiety with no significant ionic affinity. The microenvironmental effect on the binding of Cu(II) and Co(II) was quantified from solutions buffered to pH 5 with acetate ion. This effect was evident from the experimentally determined binding constants: for Cu(II) the values were 3130, 4134, 8203, 3108, and 0 N^{-1} for the IPNs with 100%, 77%, 53%, 27%, and 0% VI, respectively, while the corresponding values for Co(II) were 294, 270, 86, 129, and 0 N^{-1}, respectively. The results show that increasing the ester group content around the imidazole ligands up to an equimolar amount increases that ligand's affinity for Cu(II), after which point the affinity decreases. Conversely, the affinity of the imidazole ligand for Co(II) is low and gets lower as the ester group content increases. The ester group alone, as expected, has no affinity for either metal ion and acts only to vary the polarity within the IPN. There is, therefore, an optimum microenvironment in which imidazole's affinity for Cu(II) is maximized, probably due to an increased compatibility between the copper salt that is sorbed into the IPN and the IPN itself. The microenvironmental effect within IPNs can thus be an important means of enhancing a ligand's affinity and selectivity for a targeted metal ion.

The influence of bifunctionality on metal ion affinities was studied by forming IPNs within polystyrene beads of 4-vinylpyridine (VP), EA, and an equimolar ratio of VP:EA, followed by hydrolyzing the ester-containing IPNs to give carboxylic acid ligands [60]. The IPNs were contacted with dilute (10^{-4} N) metal ion solutions of Eu(III), Ni(II), Co(II), Zn(II), Cu(II), Hg(II), and Cd(II) in 0.10 M HNO_3. The Ni(II) distribution coefficients for the VP, acid, and VP/acid IPNs are 105, 129, and 589, respectively, indicating that a supported ligand synergistic interaction is operative. Similar results are found with Co(II) and Zn(II). Synergism is not evident with Cu(II), Hg(II), and Cd(II): for example, the Cd(II) distribution coefficients are 10, 2344, and 2399 for the VP, acid, and VP/acid IPNs, respectively, indicating that the carboxylic acid ligand is solely responsible for the sorption process. With Eu(III), another effect was identified: the distribution coefficients for the IPNs as listed above are 3, 2042, and 22, respectively. In this case, bifunctionality leads to antagonism, not synergism. As confirmed by FTIR, there is hydrogen bonding between the pyridyl and carboxylic acid moieties and Eu(III) binding is not sufficiently strong to overcome it. Ions such as Cd(II) overcome the hydrogen bonding for binding to occur to one of the ligands, while ions such as Ni(II) overcome the hydrogen bonding and bind more strongly to both ligands than to either one alone. The IPNs are therefore an effective means of defining the conditions under which both the microenvironmental effect and supported ligand synergistic interaction can be effectively utilized for greater metal ion affinities and selectivities.

VIII. BIFUNCTIONAL SOLVENT IMPREGNATED RESINS

The versatility of solvent extraction and the operational simplicity of ion exchange resins may be combined in the utilization of solvent impregnated resins (SIRs). In this technique, an organophilic complexant is sorbed within macroporous copolymer beads and the combined unit then handled as a polymer-supported reagent. Any complexant or combination of complexants may be used to form SIRs, allowing a straightforward study of possible synergistic pairings without the need to develop immobilization strategies for new bifunctional immobilized polymers. In one of many examples, a SIR was prepared by combining DEHPA with TOPO and sorbing both into a macroporous support, Amberlite XAD-2 [61]. The solution with an equimolar ratio of the two complexants allowed for better separation of Zn(II)/Cu(II) and Zn(II)/Cd(II) mixtures. A bifunctional complexant, O-methyldihexyl-phosphine-oxide O'-hexyl-2-ethyl phosphoric acid, was also sorbed within Amberlite XAD-2 [62]. Zn(II), Cu(II), and Cd(II) were all complexed to a significant extent from 0.1 M $NaNO_3$ with the extent of complexation increasing as the solution pH increased.

A new bifunctional SIR was introduced in which ion-exchange ligands were immobilized on the macroporous support, and an organophilic soluble coordinating complexant was sorbed into the polymer [63]. The key advantage of solvent extraction—flexibility in choosing among a large variety of selective soluble complexants—is retained and combined with the ease of synthesizing a monofunctional ion exchange resin. In this case, sulfonic acid ligands were immobilized onto macroporous polystyrene beads, and tetrathia-14-crown-4 was sorbed into the beads. The degree of sulfonation of the polystyrene was a critical variable, with a lower degree of functionalization providing the most favorable results due to its compatibility with the organophilic crown. The crown ether itself, dissolved in toluene or sorbed within nonfunctionalized polystyrene beads, had no affinity for Cu(II) from sulfuric acid solutions. However, the crown/sulfonated SIR showed a synergistic enhancement of up to two orders of magnitude in the amount of Cu(II) complexed from sulfuric acid solution, which is perhaps the first observation of synergism in a functionalized SIR.

An important disadvantage of SIRs is the loss through aqueous phase solubility of the sorbed complexant. This is a significant problem and precludes their adaption to preparative scale applications. In order to obviate this loss of complexant, a new SIR has been prepared wherein a thin membrane coating is formed around each bead [64]. This coating is hydrophilic, thus preventing transport of the hydrophobic complexant out of the bead while permitting transport of the hydrophilic metal ion into the bead. The method by which this was accomplished involved (1) con-

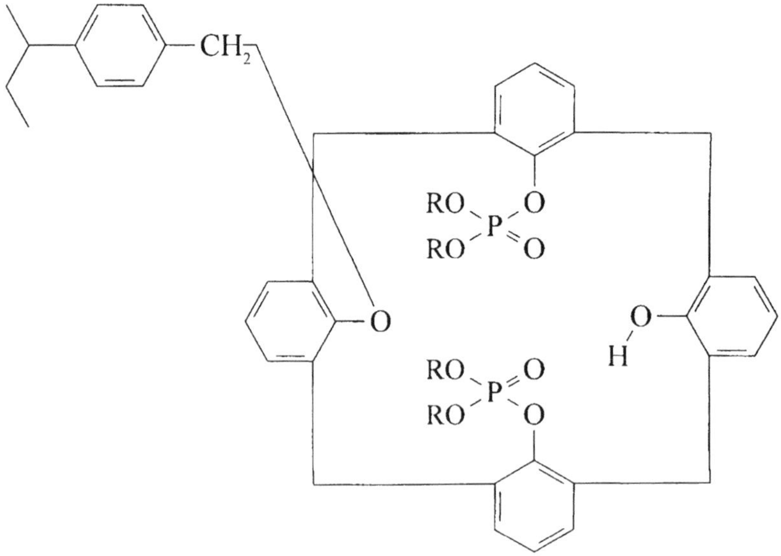

Figure 18   Polystyrene-supported calix[4]arene.

verting the $-CH_2Cl$ groups on the surface of macroporous crosslinked poly(vinylbenzyl chloride) beads to phosphonate groups through an Arbusov reaction and then to vinyl groups through a Wadsworth–Emmons reaction, (2) sorbing the complexant into the beads, and (3) forming a thin membrane coating that is anchored to the surface of the beads from a 2% aqueous solution of glycidyl methacrylate/bis(acrylamide) in a 75/25 weight ratio. After DEHPA was encapsulated in this manner, the beads were contacted with a solution of 10^{-4} N $Cu(NO_3)_2$ buffered to pH 8.4 in an ammonia/ammonium nitrate solution and found to complex 96% of the Cu(II) present. Most importantly, this level of complexation remained constant after five cycles of regeneration with 1 M HNO_3 and contact with fresh solution. By comparison, the standard SIR prepared in the same manner but without coat formation complexed the same amount of Cu(II) as the coated SIR in the first contact, but dropped to 23.2% Cu(II) after the first regeneration and 11.2% Cu(II) after the second regeneration. The hydrophilic membrane coating thus successfully retained the complexant within the pores of the beads without impeding the kinetics of metal ion transport and complex formation. The development of these *high-stability SIRs* holds significant potential for long-term applications where soluble complexants with a targeted ionic selectivity have been identified.

IX. CURRENT DEVELOPMENTS

Research in our laboratory continues with bifunctional polymers involving a wide array of ligands, including aminophosphonate, phosphonoacetate, and phosphopyridyl. As will be reported in an upcoming publication, we have also synthesized a polymer-supported calixarene [65]. The immobilized calix[4]arene has a high affinity for cesium ions from alkaline solutions and has the potential for being an important probe of intraligand cooperative effects since ion-complexing ligands can be placed around the ring at either the aromatic or phenolic ends. In one example, the calixarene ring in the cone conformation was substituted with phosphate groups at two of the phenolic sites (Fig. 18). The order of affinity was Fe(III) > Pb(II) > Cu(II) > Ni(II) with quantitative complexation of Fe(III) and Pb(II) from 0.01 M HNO_3 solution. The unsubstituted calixarene, poly (hydroxystyrene), and phosphate-substituted polystyrene had no affinity for any of the ions. It may be concluded that the high metal ion affinities and selectivities were due to a combination of the calixarene ring and the ion-complexing ligands in a conformation that permitted intraligand cooperation.

X. CONCLUSION

The design and development of ion-selective polymer-supported reagents is an important challenge for the 21st century. Bifunctional polymers in the form of beads, membranes, and high-stability SIRs are exceptionally promising in their ability to enhance our understanding of metal ion coordination chemistry as well as in their application to environmental remediation, water purification, chromatographic separations, and sensor technology. Target selectivities will be best achieved with structures that maximize intraligand cooperation, such as immobilized macrocycles. Further breakthroughs from leading laboratories around the world are anticipated.

ACKNOWLEDGMENTS

We are grateful for the long-standing support we have received from the U.S. Department of Energy, Office of Basic Energy Sciences. Additionally, we have had productive collaborations with colleagues at the Oak Ridge National Laboratory (Drs. Bruce Moyer and Gilbert Brown have been principal collaborators), Argonne National Laboratory (it was the close cooperation with Dr. Phil Horwitz that led to Diphonix), and Los Alamos National Laboratory (Dr. Gordon Jarvinen has been a valued collaborator there). The graduate students who have worked with so much enthusiasm on this research are noted as co-authors in the publications cited. A special thanks goes to Dr. Robert Ober for compiling the references used in this report.

REFERENCES

1. CA Lesburg, C Huang, DW Christianson, CA Fierke. Biochemistry 36: 15780–15791, 1997.

2. RB Thompson, BP Maliwal, CA Fierke. Anal Chem 70:1749–1754, 1998.
3. RB Thompson, BP Maliwal, CA Fierke. Anal Biochem 267:185–195, 1999.
4. JA Hunt, M Ahmed, CA Fierke. Biochemistry 38:9054–9062, 1999.
5. JA Hunt, CA Fierke. J Biol Chem 272:20364–20372, 1997.
6. FWE Strelow. Anal Chem 32:1185–1188, 1960.
7. KA Kraus, RJ Raridon. J Phys Chem 63:1901–1907, 1959.
8. MJ Slater, BH Lucas, GM Ritcey. CIM Bulletin 117–123, 1978.
9. F Helfferich. J Phys Chem 69:1178–1187, 1965.
10. GE Janauer, GO Ramseyer, JW Lin Anal Chim Acta 73:311–319, 1974.
11. TC Lo, MHI Baird, C Hanson. Handbook of Solvent Extraction. New York: Wiley, 1983.
12. T Sekine, Y Hasegawa. Solvent Extraction Chemistry: Fundamentals and Applications. New York: Marcel Dekker, 1977.
13. JM Sanchez, M Hidalgo, V Salvado, M Valiente. Solvent Extr Ion Exch 17:455–474, 1999.
14. E Antico, A Masana, M Hildago, V Salvado, M Iglesias, M Valiente. Anal Chim Acta 327:267–276, 1996.
15. RK Biswas, DA Begum. Hydrometallurgy 50:153–168, 1998.
16. PM Mapara, AG Godbole, R Swarup, NV Thakur. Hydrometallurgy 49:197–201, 1998.
17. NB Devi, KC Nathsarma, V Chakravortty. Hydrometallurgy 49:47–61, 1998.
18. BY Mishra, PM Dhadke. Sep Sci Technol 33:1681–1692, 1998.
19. AC Muscatello, SL Yarbro, SF Marsh. In: L Cecille, M Casarci, L Pietrelli, eds. New Separation Chemical Technology for Radioactive Waste and Other Specific Applications (Proceedings of Technical Seminar, Los Alamos, NM). London: Elsevier, 1991.
20. SL Blaha, DJ McCabe, RT Paine, KW Thomas. Radiochim Acta 46:123–125, 1989.
21. M Zangen. J Inorg Nucl Chem 25:581–594, 1963.
22. T Hano, M Matsumoto, T Ohtake, N Egashira, F Hori. Solvent Extr Ion Exch 10:195–206, 1992.
23. KK Sahu, RP Das. Metall Mater Trans. B 28B:181–189, 1997.
24. G Modolo, R Odoj. Solvent Extr Ion Exch 17:33–53, 1999.
25. ML Dietz, AH Bond, BP Hay, R Chiarizia, VJ Huber, AW Herlinger. Chem Commun 1177–1178, 1999.
26. SI El-Dessouky, JA Daoud, HF Aly. Radiochim Acta 85:79–82, 1999.
27. SD Alexandratos. Sep Purif Methods 17:67–102, 1988.

28. SD Alexandratos, MA Strand, DR Quillen, AJ Walder. Macromolecules 18:829–835, 1985.
29. SD Alexandratos, DL Wilson, MA Strand, DR Quillen, AJ Walder. Macromolecules 18:835–840, 1985.
30. SD Alexandratos, DL Wilson. Macromolecules 19:280–287, 1986.
31. SD Alexandratos, DL Wilson, PT Kaiser, WJ McDowell. React Polym 5: 23–35, 1987.
32. GK Schweitzer, AMBM Radzi, SD Alexandratos. Anal Chim Acta 209: 363–366, 1988.
33. SD Alexandratos, DR Quillen, WJ McDowell. Sep Sci Technol 22:983–995, 1987.
34. SD Alexandratos, PT Kaiser. Solvent Extr Ion Exch 10:539–557, 1992.
35. SD Alexandratos, DR Quillen, ME Bates. Macromolecules 20:1191–1196, 1987.
36. SD Alexandratos, ME Bates, AJ Walder, WJ McDowell. Sep Sci Technol 23:1915–1927, 1988.
37. SD Alexandratos, ME Bates. Macromolecules 21:2905–2910, 1988.
38. SD Alexandratos, DW Crick, DR Quillen. Ind Eng Chem Res 30:772–778, 1991.
39. SD Alexandratos, CA Shelley, EP Horwitz, R Chiarizia. Solvent Extr Ion Exch 16:951–966, 1998.
40. SD Alexandratos, S Natesan. Eur Polym J 35:431–436, 1999.
41. SD Alexandratos, LA Hussain. Ind Eng Chem Res 34:251–254, 1995.
42. MA Hines, JC Sullivan, KL Nash. Inorg Chem 32:1820–1823, 1993.
43. SD Alexandratos, AW Trochimczuk, DW Crick, EP Horwitz, RC Gatrone, R Chiarizia. Macromolecules 29:1021–1026, 1996.
44. R Chiarizia, EP Horwitz, SD Alexandratos, MJ Gula. Sep Sci Technol 32: 1–35, 1997.
45. R Chiarizia, EP Horwitz, SD Alexandratos. Solvent Extr Ion Exch 12:211–237, 1994.
46. EP Horwitz, R Chiarizia, H Diamond, RC Gatrone, SD Alexandratos, AW Trochimczuk, DW Crick. Solvent Extr Ion Exch 11:943–966, 1993.
47. EP Horwitz, R Chiarizia, SD Alexandratos. Solvent Extr Ion Exch 12:831–845, 1994.
48. R Chiarizia, EP Horwitz, RC Gatrone, SD Alexandratos, AW Trochimczuk, DW Crick. Solvent Extr Ion Exch 11:967–985, 1993.
49. AW Trochimczuk, EP Horwitz, SD Alexandratos. Sep Sci Technol 29: 543–549, 1994.
50. SD Alexandratos, AW Trochimczuk, EP Horwitz, RC Gatrone. J Appl Polym Sci 61:273–278, 1996.

51. R Chiarizia, KA D'Arcy, EP Horwitz, SD Alexandratos, AW Trochimczuk. Solvent Extr Ion Exch 14:519–542, 1996.
52. R Chiarizia, EP Horwitz, KA D'Arcy, SD Alexandratos, AW Trochimczuk. Solvent Extr Ion Exch 14:1077–1100, 1996.
53. R Chiarizia, EP Horwitz, RA Beauvais, SD Alexandratos. Solvent Extr Ion Exch 16:875–898, 1998.
54. SD Alexandratos, LA Hussain. Macromolecules 31:3235–3238, 1998.
55. PV Bonnesen, GM Brown, LB Bavoux, DJ Presley, BA Moyer, SD Alexandratos, V Patel, R Ober. Environ Sci Technol 34:3761–3766, 2000.
56. RG Pearson. J Am Chem Soc 85:3533, 1963.
57. B Gu, GM Brown, PV Bonnesen, L Liang, BA Moyer, R Ober, SD Alexandratos. Environ Sci Technol 34:1075–1080, 2000.
58. LH Sperling. Interpenetrating Polymer Networks and Related Materials. New York: Plenum, 1981.
59. SD Alexandratos, CG Ciaccio, R Beauvais. React Polym 19:137–143, 1993.
60. AA Kriger, BA Moyer, SD Alexandratos. React Polym 24:35–39, 1994.
61. JL Cortina, N Miralles, AM Sastre, M Aguilar. React Funct Polym 32: 221–229, 1997.
62. JL Cortina, N Miralles, M Aguilar, A Warshawsky. React Funct Polym 27:61–73, 1995.
63. BA Moyer, GN Case, SD Alexandratos, AA Kriger. Anal Chem 65:3389–3395, 1993.
64. SD Alexandratos, KP Ripperger. Ind Eng Chem Res 37:4756–4760, 1998.
65. SD Alexandratos, S Natesan. Macromolecules (submitted, 2000).

2

Recovery of Valuable Species from Dissolving Solids Using Ion Exchange

Jannie S. J. van Deventer

University of Melbourne, Victoria, Australia

P. G. R. de Villiers and L. Lorenzen

University of Stellenbosch, Stellenbosch, South Africa

Ion exchange resins can be used to recover valuable species from sparingly soluble ores or solid particles that slightly dissociate to give traces of the valuable ions in solution. A dissolution equilibrium is established between the dissolved ions in solution and the solid particles. If the traces of dissolved ions are removed from solution by ion exchange, this equilibrium is displaced so that further dissolution takes place according to Le Chatelier's principle. It is essential that the ion exchange resin has a high equilibrium loading of the valuable species at ultra-low equilibrium concentrations of the solution. The ion exchange resin should also have a high degree of selectivity for the valuable species. Complete dissolution of the sparingly soluble solid, hence complete liberation of the valuable species, can be achieved if a sufficient amount of ion exchanger is present. Improved extraction of the valuable species can be obtained if the associated contaminant species are removed from the solution phase during leaching. Precipitation and weak acid formation reactions between the contaminant and the exchanged counter ion, as well as mixtures of cation and anion exchange resins, can be used to reduce the concentration of the contaminants. Enhanced recovery of the valuable species is obtained by increasing the solubility of the solids through the creation of electrolyte solutions from the exchanged counterions initially saturating a mixture of cation and anion exchange resins. The equilibrium relationships for this simultaneous leaching and sorption process are outlined in this chapter. The selective sorption of calcium and magnesium from a sparingly soluble dolomitic ore as well as the recovery of lead from lead sulfate were used as case studies to demonstrate the principles of the process.

I. INTRODUCTION

Many resource industries have become marginal owing to declining ore grades, high operating costs, and unstable markets. Some mineral deposits have remained economical as a result of the extraction of the by-products rather than the primary metal. In many marginal hydrometallurgical operations it is not technically or economically feasible to extract all valuable species from process streams owing to their low concentrations in either the solids or the leachate. Except in the case of high value species such as gold and the platinum group metals, it is not possible to recover trace components from fresh ore or tailings using existing technology. For example, rare earths are sometimes available at levels below 0.05%, which is too low judged by modern standards of hydrometallurgical recovery. This is often aggravated by complex mineralogy, which inhibits efficient extraction of the particular mineral species without liberating contaminants (unwanted species) associated with the valuable mineral.

Renewed interest in the development of technology for extraction of species from solids or solutions of ultra-low concentration is not limited to the hydrometallurgical recovery of metals from primary ores. The recovery of metals from "artificial" ores can also be considered: for example, the recovery of nickel from spent catalysts used by the chemical industry, the extraction of cobalt, molybdenum, and gold from fly ash generated by power stations, and the recovery of gold from recycled electronic circuit boards. Furthermore, there is an increasing need to remove trace contaminants from waste soil, sludge, or dumps during environmental remediation. The economic viability of such a process depends largely on the technology employed, which needs to be highly selective and energy efficient in order to compensate for the low concentration of valuable species. Fortunately, recent advances made in ion exchange technology and the synthesis of highly selective inorganic adsorbents are causing a paradigm shift in hydrometallurgical technology. A high degree of artificially induced selectivity for the valuable species, combined with properties such as ultra-high capacities and abrasion resistance in slurries, makes ion exchange resins and modern inorganic adsorbents suitable extractants for new processes involving simultaneous dissolution and adsorption. Furthermore, an extremely low solution phase concentration of the valuable species is adequate to effect mass transfer that will result in significantly high loadings of the valuable species on the adsorbent. It is unfortunate that the literature gives almost no information about the relationship between the nature of functional groups and the shape of the adsorption isotherm for either ion exchange resins or inorganic adsorbents. Although it is not the focus of this chapter, an understanding of this relationship is pivotal to the success of a combined dissolution and adsorption technology.

In this chapter modern strategies of hydrometallurgical extraction will be outlined. The use of selective ion exchange resins in simultaneous dissolution and adsorption processes will be explained with reference to extraction from sparingly soluble solids. It will be explained that minimal consumption or complete absence of leaching reagents is possible if highly selective ion exchange resins are used to shift the equilibrium of dissolution. A sparingly soluble dolomitic ore and lead sulfate will be used as model solids in the case studies outlined.

II. HYDROMETALLURGICAL EXTRACTION

Leaching, which is the liberation of metals from ores by chemical dissolution, forms the basis of most hydrometallurgical extraction processes. The principal aim of leaching is to liberate the maximum amount of the valuable species as selectively as possible. A compromise exists between the recovery and the grade of the valuable species in the leachate, with the recovery usually being the dominant driving force. The leaching procedure is typically aggressive, especially in the case of refractory ores, in that contaminants (unwanted species) associated with the valuable species are liberated. Nevertheless, leaching is an established and relatively successful method of mineral extraction, especially when treating high grade ores. In contrast, the leaching of low grade ore is often uneconomical owing to high consumption of the leaching agent and the resultant leachate containing too low a concentration of the valuable species relative to the contaminants. This implies that large volumes of this pregnant solution must be processed in downstream unit operations in order to obtain a sufficiently high concentration of valuable species for final recovery. In some cases the leaching of valuable species from a host ore is concentration dependent, which means that it becomes increasingly difficult to transfer valuable species from the ore into solution as the concentration of the species in solution increases, in accordance with Le Chatelier's principle. To counteract this phenomenon, multistage leaching circuits are required, preferably operating in a countercurrent fashion with an adsorbent present.

Consequently, there is a need to develop a technology that makes the extractive process more selective with respect to the valuable species and hence minimizes the size of downstream processing units. Moreover, the liberated valuable species should be concentrated in an extractive phase that can be separated easily from the leached slurry of ore particles. It is also important to minimize the consumption of reagents, especially by the waste components in the ore.

Ion exchange is one of the most promising new technologies available to treat low grade ores. Although the potential benefits of resin technology in the extractive industries have been recognized for a long time, progress has been inhibited by the unavailability of resins with controlled properties such as capacity, selectivity, and stability. Today it is possible to synthesize highly selective resins that are chemically and physically stable. This allows the use of ion exchange resins for use in hydrometallurgical processes such as extraction from primary ores, recovery of valuable species from recycled material, and the removal of contaminants from waste in environmental cleanup operations. The next section will explain the principles of the resin-in-pulp (RIP)/resin-in-leach (RIL) process and how the leaching and sorption processes are affected by the nature of the functional groups on the resin [1,2]. Unfortunately, these processes have not been commercialized widely and it is hoped that this chapter will aid in the understanding and acceptance of this technology. An industrial example of a process related to this technology is the extraction of rare earths from low grade kaolinite ore by percolation leaching in which recovery takes place by shifting the solid/liquid dissociation equilibrium condition.

III. PRINCIPLES OF THE RESIN-IN-LEACH PROCESS

Terms such as "insoluble" and "sparingly soluble" are encountered in the literature on hydrometallurgy, and need to be defined for the purposes of this chapter. A solid substance is defined as insoluble, or sparingly soluble, when its thermodynamic dissolution constant (called the solubility product) has a value of 10^{-5} or less. Equation (1) is a general representation of the dissociation (dissolution) reaction of substance $B_{v_B}Y_{v_Y}$ into v_B moles of its constituent cation, B^{z_B}, and v_Y moles of its anion, Y^{z_Y}:

$$[B_{v_B}Y_{v_Y}]_{solid} \overset{K_{sp}}{\longleftrightarrow} v_B B^{z_B} + v_Y Y^{z_Y} \tag{1}$$

where the solubility product, K_{sp}, for the reaction is defined by

$$K_{sp} = [a_B]^{v_B}[a_Y]^{v_Y} = [C_B\gamma_B]^{v_B}[C_Y\gamma_Y]^{v_Y} \tag{2}$$

with C_i, a_i, and γ_i the solution phase concentration, activity, and activity coefficient of species i, respectively. Species B^{z_B} represents a valuable species, which is associated with contaminant species Y^{z_Y} (unwanted species), present as a sparingly soluble substance, $B_{v_B}Y_{v_Y}$, in an ore. The supernatant solution in equilibrium with the ore particles contains both species B^{z_B} and Y^{z_Y} at trace level concentrations [small K_{sp} value for Eq. (2)], which are considered as zero for all practical hydrometallurgical purposes. In order to recover species B^{z_B}, it must be leached from the ore particles into solution. The pregnant solution is treated by downstream unit operations that separate and recover species B^{z_B} from the solution phase containing the contaminants (species Y^{z_Y}).

During the leaching procedure, the solution phase concentration of species B^{z_B} increases dramatically. Unfortunately, the solution phase concentration of the contaminants increases simultaneously, and according to Le Chatelier's principle Eq. (1) will be shifted to the left at some total solution concentration of species B^{z_B} and Y^{z_Y}. To counteract this re-precipitation reaction, more leaching reagent is required to "pull" species B^{z_B} back into solution, which results in higher consumption of the reagent per unit mass of species B^{z_B} leached. This decreases the effective use of the leaching reagent, which increases the operational costs of the process. To counteract higher reagent consumption, multistage leaching circuits are used, where additional leaching reagent (containing no B^{z_B} or Y^{z_Y} ions) is added to the partially leached slurry. The dissociation equilibrium that exists between the solid, $B_{v_B}Y_{v_Y}$, and the dissolved constituent ions, B^{z_B} and Y^{z_Y}, is disturbed when the pregnant solution is removed between different leaching stages and is re-established by the further dissolution of the solid according to Le Chatelier's principle. It is evident that downstream processing and operating costs could be minimized if dissolution (leaching) takes place simultaneously with adsorption (removal from solution phase) of the valuable species B^{z_B} onto a separate extractive phase such as an ion exchange resin.

Valuable species can be recovered from sparingly soluble solids that slightly dissociate to give traces of the valuable ions in solution using ion exchange resins in a slurry mixture, called a resin-in-leach mixture, containing the ore body and the ion exchange resin. The traces of valuable ions B^{z_B} are removed from the solution by the ion exchange reaction, which disturbs the solid/liquid dissociation equilibrium condition. Further dissolution of the solid is required (Le Chatelier's principle) to restore the equilibrium concentration of the valuable ions B^{z_B} in solution, which are again removed by the ion exchange reaction. Complete dissolution of the solid $B_{v_B}Y_{v_Y}$, and hence complete liberation of the valuable species, can therefore be achieved by continually displacing the solid/liquid equilibrium condition, provided that an excess of the ion exchanger is added. It is evident that the shifting of equilibrium conditions is the principal mechanism in the recovery of valuable species from sparingly soluble solids in a RIL mixture.

In principle, the ion exchange process is governed by electrostatic interactions that exist between the different species; these interactions migrate between the inside of the ion exchanger phase and the external bulk solution. In order to maintain electrical neutrality in both of these phases, the ion exchange reaction must involve the transfer of stoichiometrically equivalent amounts of similarly charged species. This means that the species B^{z_B} can only migrate into the ion exchanger phase if a stoichiometrically equivalent amount of cation species A^{z_A} simultaneously migrates from the inside of the ion exchanger into the external bulk solution. The formation of species A^{z_A}, which is stoichiometrically expelled from the resin phase, depends largely on the chemical nature of the ionogenic functional groups and the interactions with the valuable species.

The functional groups (where the chemical exchange reaction occurs) of the ion exchanger may initially be saturated with species A^{z_A}, in which case the process is a straightforward example of a stoichiometrically equivalent ion exchange reaction between species B^{z_B} and A^{z_A}, which are called the counterions of the ion exchange reaction. This type of ion exchange reaction is usually encountered with what is commonly known as strong acid cation or strong base anion exchange resins. Species A^{z_A} may also be formed in a completely different way than in the above-mentioned example.

Initially, species A^{z_A} may have been part of the functional groups inside the resin, before being "activated" into participating in the ion exchange reaction by fluctuating conditions in the chemical composition of the external bulk solution. In the activated form, species A^{z_A} can participate in a stoichiometrically equivalent ion exchange reaction with species B^{z_B} provided that specific conditions predominate in the external solution phase. Ion exchange reactions of this nature are encountered with weak acid cation and weak base anion exchangers, where the solution phase pH is the activator.

Another type of ion exchange reaction occurs when species A^{z_A} is initially part of species B^{z_B}. Species A^{z_A} may be a chemical product formed during the ion exchange reaction between species B^{z_B} and the functional groups of the ion exchanger, implying that only a part of the original species B^{z_B} is permanently adsorbed by the ion exchanger. Ion exchange reactions exhibiting behavior of this nature are usually associated with what is known as chelating resins. As a result of the ion exchange reaction, electrolyte solutions are created containing species A^{z_A} and Y^{z_Y}, which may change the solubility of the sparingly soluble solid and consequently also the equilibrium conditions between the phases. The electrolyte solution can either enhance or lower the solubility of the sparingly soluble solid, leading to new equilibrium conditions.

Qualitative descriptions of processes related to this RIL process can be found in the literature. Nishino [3] studied the dissolution of Portland cement powder in an aqueous suspension containing a strongly acidic cation exchange resin in the hydrogen form. It was concluded that complete dissolution of the Portland cement could be achieved by the addition of the resin to an amount of about ten times the weight of the cement. Yang et al. [4] studied the recovery of sorbic acid from dilute aqueous suspensions with the use of an ion exchange resin containing trialkyl phosphine oxide (TRPO). The sorbic acid has a dissociation constant of 1.73×10^{-5} kmol/m^3 at 298 K, which makes it a borderline case according to the definition of a sparingly soluble solid ($K_{sp} \leq 10^{-5}$). Cloete and Marais [5] studied the recovery of acetic acid from dilute 1% aqueous solutions, using a weakly basic anion exchange resin, Duolite A-375, in the free-base form. To minimize chemical consumption of the elution reagent (OH$^-$ ions), the sorbed acetic acid was eluted from the anion resin with a slurry of lime [Ca(OH)$_2$], which resulted in the formation of calcium acetate. This solution can then be evaporated to crystallize calcium acetate or reacted with sulphuric acid to form acetic acid and gypsum, which results in product solutions of 15–20% acetic acid. The same principle was used by Kadlec et al. [6] in the regeneration of cation exchange resins used in the production of deionized water. Mild solutions of sulfuric acid were used to elute Ca^{2+} ions from the cation exchange resin in the regeneration circuit, which precipitated as CaSO$_4$.

It is evident from these examples that the RIL process can be a most useful technique for the recovery of valuable species from sparingly soluble solids, where the valuable species are present at very low concentrations. The overall extraction capability of the process depends on the solubility of the valuable species of interest and the selectivity of the ion exchange resin for the valuable species in a multicomponent solution.

IV. EQUILIBRIUM RELATIONSHIPS

The chemical composition of the RIL solution phase changes during movement to equilibrium through the ion exchange reaction, with the introduction of new chemical species that have been restricted initially to the resin phase. This shift in the solution phase composition can alter the solubility of the sparingly soluble solid, which is a function of the solubility product constant defined in Eq. (1). The solubility of the valuable species B^{z_B} can be expressed as a function of the solubility product constant and the solution phase activity coefficients, as represented in Eq. (3):

$$S = \left(\frac{K_{sp}}{[\nu_B \gamma_B]^{\nu_B} [\nu_Y \gamma_Y]^{\nu_Y}} \right)^{1/\nu_B + \nu_Y} \tag{3}$$

It is evident that the solubility S of the valuable species must change with variations in its solution phase activity coefficient, γ_B, since the solubility product, K_{sp}, is constant for a given set of physical conditions such as temperature. The shift in electrolyte composition of the RIL solution phase with the introduction of alien species will inevitably change the solution phase activity coefficient of the valuable species. Natural multicomponent electrolyte solutions behave nonideally, the extent to which is given by the deviation in the value of the solution phase activity coefficient from unity. A decrease in the value of the activity coefficient of the valuable species in Eq. (3) corresponds to a larger value of S, which relates to an increase in the solution phase concentration of the valuable species. This increase provides an additional mass transfer driving force for the sorption of the valuable species by the ion exchanger.

However, an increase in the solution phase concentration of the valuable species B^{z_B} is accompanied by an increase in the solution phase concentration of the contaminant Y^{z_Y}, which according to Le Chatelier's principle will shift the dissolution reaction, Eq. (1), to the left. This shift in the equilibrium condition, activated by an increase in the concentration of the ion common to the equilibrium reaction, is called the *common ion effect* [7]. The decline in the solubility of the valuable species due to the common ion effect is represented by Eq. (4), where the solution phase concentration of species Y^{z_Y} has increased with m moles/litre:

$$K_{sp} = [\nu_B S \gamma_B]^{\nu_B} [(\nu_Y S + m) \gamma_Y]^{\nu_Y} \tag{4}$$

There are many possibilities to compensate for the common ion effect, which in principle involves the chemical transformation of the contaminant species Y^{z_Y} to disguise it from the dissolution reaction involving the valuable species. This transformation can be effected with the use of a proper counterion species, A^{z_A}, to initially saturate the ion exchange resin. If the contaminant species Y^{z_Y} is the conjugate base of a weak acid, the solubility of the valuable species can be significantly enhanced if the cation exchange resin is initially used in the hydrogen ionic form. The exchanged hydrogen ions will form a weak acid with the contaminant Y^{z_Y}, as represented by Eq. (5):

$$[(B^{z_B})_{v_B}(Y^{z_Y})_{v_Y}]_s + v_Y H^+ \overset{K_a}{\leftrightarrow} v_B B^{z_B} + v_Y HY^{(z_Y+1)} \qquad (5)$$

with the enhanced solubility product constant given by

$$K_a = \frac{(a_B)^{v_B}(a_{HY})^{v_Y}}{(a_H)^{v_Y}} = \frac{K_{sp}}{(K_a)^{v_a}} \qquad (6)$$

It is evident that the smaller the value is of the dissociation constant of the weak acid, the larger the value will be of the enhanced solubility product.

Another possibility to counteract the common ion effect involves the precipitation of the contaminant species Y^{z_Y} with the exchanged counterion species A^{z_A}. This option will only be viable if electrolyte sorption by the ion exchanger is virtually nonexistent, especially toward the contaminant Y^{z_Y}. The obvious reason for this is the possibility of precipitation within the ion exchanger pores, which will severely retard the sorption kinetics. Complete Donnan-exclusion of electrolytes holds extremely well for gel-type resins of moderate to high degrees of crosslinking, in which case precipitation can be used. In the case of macroporous ion exchange resins, electrolyte sorption can be as much as 2% of the total ionic species content inside the resin phase. If the pore liquid concentration of contaminant Y^{z_Y} is sufficiently high, precipitation may be triggered that renders this possibility unfeasible.

The solubility of the valuable species can be enhanced significantly with the use of RIL slurries employing mixtures of cation and anion exchange resins. An optimum value exists for the ratio of the amounts of each type of resin. To promote this effect the value of m in Eq. (4) must be minimized. Therefore, the distribution of the different species between the three phases (ore/solution/resin) for a multicomponent mixture must be predicted.

A general expression for the sorption of the valuable species B^{z_B} by the ion exchanger initially saturated with species A^{z_A} is

$$z_B[(R^-)_{z_A} A^{z_A}] + z_A B^{z_B} \overset{K_A^B}{\leftrightarrow} z_A[(R^-)_{z_B} B^{z_B}] + z_B A^{z_A} \qquad (7)$$

Equation (7) reduces to different equations depending on the values of the valencies of the species z_A and z_B. The thermodynamic equilibrium constant [8–11] for the ion exchange reaction is given by

$$K_A^B = \frac{(y_B \overline{\gamma_B})^{z_A} (C_A \gamma_A)^{z_B}}{(y_A \overline{\gamma_A})^{z_B} (C_B \gamma_B)^{z_A}} \qquad (8)$$

where y_i and C_i represent the resin phase mole fraction and solution phase concentration of species, i, respectively. The thermodynamic dissolution coefficient [8] can be derived from a stoichiometric combination of Eqs. (1) and (7) for the particular values of the species valences. The derivations are straightforward and are summarized in Table 1, which gives the thermodynamic dissolution coefficient obtained for different $|Z_B|$ and $|Z_A|$ relations.

A mass balance can be performed for each of the species to obtain its equilibrium distribution between the different phases for a batch reactor. It is assumed that enough of the sparingly soluble solid is present to prevent complete dissolution by the ion exchange resin and that no sorption of electrolyte occurs. It is also assumed that species B^{z_B} and Y^{z_Y} are initially present only in the ore phase, with species A^{z_A} initially present only in the resin phase. Initially, the solution contains no added electrolyte species. It is also assumed that no precipitation reactions occur.

At equilibrium, the solution phase concentrations of the species B^{z_B}, Y^{z_Y}, and A^{z_A} are given by Eqs. (9), (10), and (11), respectively:

$$C_B = \frac{|z_Y| C_Y - |z_A| C_A}{|z_B|} \qquad (9)$$

$$C_Y = v_Y S \qquad (10)$$

$$C_A = \frac{Q(1 - \varepsilon) DPRV\, y_B}{|z_A| V} \qquad (11)$$

Table 1 Thermodynamic Dissolution Coefficients for Different RIL Slurries Containing Either a Cation or an Anion Exchange Resin

Dissolution reaction	Cation exchange reaction	K_{sp}	K_A^B	K_{diss}
$|z_B| = |z_Y|$	$|z_B| > |z_A|$, $|z_B|/|z_A| =$ integer	$(a_B)(a_Y)$	$\dfrac{(\overline{a_B})(a_A)^{z_B/z_A}}{(\overline{a_A})^{a_B/z_A}(a_B)}$	$K_{sp}\,K_A^B$
$|z_B| = |z_Y|$	$|z_B| > |z_A|$, $|z_B|/|z_A| \neq$ integer or $|z_A| > |z_B|$, $|z_A|/|z_B| \neq$ integer	$(a_B)(a_Y)$	$\dfrac{(\overline{a_B})^{z_A}(a_A)^{z_B}}{(\overline{a_A})^{z_B}(a_B)^{z_A}}$	$[K_{sp}]^{z_A}K_A^B$
$|z_B| = |z_Y|$	$|z_A| > |z_B|$, $|z_A|/|z_B| =$ integer	$(a_B)(a_Y)$	$\dfrac{(\overline{a_B})^{z_A/z_B}(a_A)}{(\overline{a_A})(a_B)^{z_A/z_B}}$	$[K_{sp}]^{z_A/z_B}K_A^B$
$|z_B| = |z_Y|$	$|z_B| = |z_A|$	$(a_B)(a_Y)$	$\dfrac{(\overline{a_B})(a_A)}{(\overline{a_A})(a_B)}$	$K_{sp}K_A^B$
$|z_B| \neq |z_Y|$	$|z_B| = |z_A|$	$(a_B)^{z_Y}(a_Y)^{z_B}$	$\dfrac{(\overline{a_B})(a_A)}{(\overline{a_A})(a_B)}$	$K_{sp}[K_A^B]^{z_Y}$
$|z_B| \neq |z_Y|$	$|z_B| > |z_A|$, $|z_B|/|z_A| =$ integer	$(a_B)^{z_Y}(a_Y)^{z_B}$	$\dfrac{(\overline{a_B})(a_A)^{z_B z_A}}{(\overline{a_A})^{z_B/z_A}(a_B)}$	$K_{sp}[K_A^B]^{z_Y}$
$|z_B| \neq |z_Y|$	$|z_B| > |z_A|$, $|z_B|/|z_A| \neq$ integer or $|z_A| > |z_B|$, $|z_A|/|z_B| \neq$ integer	$(a_B)^{z_Y}(a_Y)^{z_B}$	$\dfrac{(\overline{a_B})^{z_A}(a_A)^{z_B}}{(\overline{a_A})^{z_B}(a_B)^{z_A}}$	$[K_{sp}]^{z_A}[K_A^B]^{z_Y}$
$|z_B| \neq |z_Y|$	$|z_A| > |z_B|$, $|z_A|/|z_B| =$ integer	$(a_B)^{z_Y}(a_Y)^{z_B}$	$\dfrac{(\overline{a_B})^{z_A/z_B}(a_A)}{(\overline{a_A})(a_B)^{z_A/z_B}}$	$[K_{sp}]^{z_A/z_B}[K_A^B]^{z_Y}$

Source: Ref. 2.

The value of y_B, required in Eq. (11), can be obtained from the expression for the thermodynamic equilibrium constant as a function of the solution phase concentration of the contaminant C_Y, given by

$$K_A^B = \frac{(y\,\overline{\gamma_B})^{z_A}\left(\dfrac{Q(1-\varepsilon)\,DPRV}{V|z_A|}y_B\gamma_A\right)^{z_B}}{[(1-y_B)\overline{\gamma_A}]^{z_B}\left[\left(\dfrac{|z_Y|}{|z_B|}C_Y - \dfrac{Q(1-\varepsilon)\,DPRV}{V|z_B|}y_B\right)\gamma_B\right]^{z_A}}$$

(12)

which is also a function of the solution volume, V, the volume of resin used, $DPRV$, the resin capacity, Q, and both the solution and resin phase activity coefficients of the different species. From Eq. (10) it is evident that the value of the affected solubility, S, is required, which can be obtained from the value of m, given by Eq. (4) and

$$m = \frac{|z_A|\,C_A}{|z_Y|}$$

(13)

Equations (9) to (13) must be solved simultaneously with appropriate models to predict the resin and solution phase activity coefficients of the species. Numerous thermodynamic models are available in the literature to predict solution phase activity coefficients. Pitzer's model gives accurate results for a wide range of electrolyte species [12,13]. Wilson's model can be used to predict the resin phase coefficients [14].

V. ENHANCED DISSOLUTION USING DOLOMITE AS A CASE STUDY

In this chapter lead sulfate and a dolomitic ore are used as models of sparingly soluble solids. Dolomite is a combination of calcium carbonate ($CaCO_3$) and magnesium carbonate ($MgCO_3$) of low solubility. In this section dolomite is used as an example of how dissolution could be enhanced by the formation of a precipitate or weak acid (or weak base) between the counterion initially on the ion exchange resin and the co-ion from the sparingly soluble solid. Magnesium is taken as the valuable species and calcium as the unwanted waste species.

A. Formation of a Precipitate

The dissolution of the insoluble solid ($CaCO_3$) can further be enhanced if some precipitate, for example $Sr\,CO_3$, is formed between the counterion that initially saturates the ion exchange resin [counterion A in Eq. (7) representing strontium, Sr^{2+}] and the co-ion initially in the solid [co-ion Y in Eq. (1) representing CO_3^{2-}]. The dissolution process, as presented by Eqs. (1) and (7), results in traces of the co-ion Y and counterion A being present in the solution due to the dissolution of the valuable ion B (Ca^{2+}) from the solid, which is exchanged for ion A (Sr^{2+}). Counterion A and co-ion Y can then form a precipitate according to the following equation:

$$z_Y(A^{z_A}) + z_A(Y^{z_Y}) \leftrightarrow \left[(A^{z_A})_{z_Y}(Y^{z_Y})_{z_A}\right]_S \tag{14}$$

with solubility product given by

$$K_{sp2} = (a_A)^{z_Y}(a_Y)^{z_A} = (C_{Sr^{2+}})^2(C_{CO_3^{2-}})^2 \tag{15}$$

Equations (1), (7), and (14) could be combined stoichiometrically to give the complete dissolution reaction for a variety of cases. If the precipitate formed is very insoluble, its solubility product in Eq. (15) will have a very small value, which will increase the overall equilibrium dissolution constant and thereby enhance the extraction of the counterion B from the sparingly soluble solid. If a mixed bed of ion exchangers is used, care should be taken before selecting the initial resin states for the ion exchangers since the formation of a precipitate with the valuable species will significantly decrease the recovery.

When the ion exchangers are added to the slurry containing the valuable magnesium, the small traces of Mg^{2+} and CO_3^{2-} ions will be exchanged for H^+ and OH^- ions, respectively. This is the first dissolution step in the overall mechanism. The exchanged OH^- ions will combine with the sufficient traces of Mg^{2+} ions present in the solution and precipitate due to the fact that the solubility product of $Mg(OH)_2$ is 10^6 times smaller than that of $MgCO_3$. This precipitation reaction is the second step in the overall dissolution mechanism. Since the solution is now largely robbed of its magnesium traces, more of the original particles will dissolve to form new traces of magnesium and carbonate ions. These new traces of ions are exchanged for the counterions initially saturating the ion exchange resins, resulting in new H^+ and OH^- ions being introduced into the solution which will form more of the new precipitate. This process of magnesium dissolution from the original ore via ion exchange and precipitation will continue until the $Mg(OH)_2$ precipitate that formed starts dissolving. This is the secondary dissolution step of the overall dissolution mechanism.

It is also possible to have a mixed bed ion exchange configuration, but without the formation of the secondary precipitate. For example, in the recovery of calcium from a calcite ore ($CaCO_3$, $K_{sp} = 9.29 \times 10^{-9}$ at 20°C [15]), the $Ca(OH)_2$ precipitate will never form due to the fact that the solubility product for $Ca(OH)_2$ ($K_{sp} = 4.42 \times 10^{-5}$ at 20°C [15,16]) is larger than that of $CaCO_3$. The Ca^{2+} ions will be exchanged for H^+ ions and the CO_3^{2-} ions for OH^- ions, thus replacing the calcite with water formed by the combination of the exchanged H^+ and OH^- ions.

B. Formation of a Weak Acid

If the co-ion initially in the insoluble solid forms a weak acid with the counterion initially present on the ion exchange resin, the dissolution of the solid can further be enhanced. There are many weak acids of great chemical and physiological importance that can significantly enhance the dissolution with ion exchange reactions. Almost all of the valuable cation species in insoluble ores are accompanied by anions that can form a range of different polyprotic or monoprotic weak (or strong) acids with the counterion initially saturating the ion exchange resin. The weaker the acid that forms, the smaller will be its dissociation constants. It is this small dissociation constant of the weak acid that enhances further dissolution of the sparingly soluble ore. This is achieved by removing the free ionic species in solution by forming molecules or larger ions that do not interfere with the ion exchange process as strongly (the extent to which depends on the polarizability of the molecule and the charge of the ions), as is the case with free and more mobile ionic species.

The dissociation of a weak polyprotic acid is given by

First dissociation step:

$$H_{z_Y} Y \leftrightarrow H^+ + H_{(z_Y-1)} Y^-$$

First dissociation constant:

$$K_1 = \frac{[H^+] \, [H_{(z_Y-1)}Y^-]}{[H_{z_Y}Y]}$$

z_Yth dissociation step:

$$HY^{(z_Y+1)} \leftrightarrow H^+ + Y^{z_Y}$$

z_Yth dissociation constant:

$$K_{z_Y} = \frac{[H^+] \, [Y^{z_Y}]}{[HY^{(z_Y+1)}]} \tag{16}$$

The first ionization step of a polyprotic acid is stronger than the second, which is stronger than the third. An example is the ionization of carbonic acid (H_2CO_3), which is a diprotic acid with first dissociation constant $K_1 = 3.5 \times 10^{-7}$ and second dissociation constant $K_2 = 6.0 \times 10^{-11}$ [7]. K_1 is much larger than K_2 by a factor of between 5000 and 6000. The dissociation of H_2CO_3 is the primary source of hydrogen ions in the solution, with the dissociation of HCO_3^- being negligible. Therefore, if any hydrogen ions are present in a solution with free carbonate ions, it will almost completely combine to form HCO_3^- at first and H_2CO_3 with further protonation according to the reverse of dissociation steps of H_2CO_3.

VI. EXPERIMENTAL METHODS

A. Model Solids

It was the aim of this study to investigate the simultaneous sorption and dissolution equilibria that exist in an aqueous suspension of a sparingly soluble solid containing an ion exchange resin to form a resin-in-pulp system. For the purpose of this study two model solids with low solubility were selected. First, a slurry of powdered pure lead sulfate ($PbSO_4$), with particle size 100% -75 μm and $K_{sp} = 1.6 \times 10^{-10}$ at 25° C, was contacted with anion and cation exchange resins in different initial ionic forms. The second set of experiments used dolomite ore from the Western Cape in South Africa, consisting mainly of calcium and magnesium carbonate (29.5% CaO, 21.1% MgO, and 2.8% acid insolubles by mass). The average values for the solubility products of calcium and magnesium in this material were 4.12×10^{-8} and 3.42×10^{-8}, respectively. The behavior of this natural material in terms of the relative dissolution of calcium and magnesium in the presence of anion and cation exchange resins will be outlined subsequently. In further work the slurries of sparingly soluble solids were contacted with mixtures of cation and anion exchange resins. This was done in order to determine the extent to which the solubility of the solids and the equilibrium resin loading were affected by the different electrolytes formed during ion exchange.

B. Ion Exchange Resins

Industrial grade Duolite resins (manufactured by Rohm & Haas Co., supplied by ACIX/NCP Ltd., Germiston, South Africa) were used in all the experiments. In some experiments a macroreticular-type strong acid cation exchange resin, Duolite C26, with sulfonic functional groups and an exchange capacity of 1.8–2.0 equivalent moles per liter (eq mol/L) of fwsv (free wet settled volume) resin was used. In selected experiments on the dolomitic material a gel-type strong acid cation exchange resin, Duolite C20, with sulfonic functional groups and an exchange capacity of 2.1–2.2 eq mol/L fwsv was used. The anion exchange resin used was Duolite A161, a macroreticular-type strong base anion exchange resin with quaternary ammonium functional groups, with an exchange capacity of 1.0 eq mol/L fwsv.

A newly opened resin, as received from the suppliers, is hardly fit for any experimental procedure in which the aim is to obtain accurate and reproducible results. The resin particles must be carefully selected since the rates of ion exchange and sorption equilibria depend strongly on physical properties such as particle shape, size, porosity, and chemical composition. In this study the resin was first thoroughly washed with distilled water to remove any excess electrolyte solution or contaminants left over from the manufacturing process. The resin was then screened into distinct size fractions. The size fraction 711–900 μm was used in experiments using lead sulfate as a model solid, while the size range 850–1180 μm was used in experiments on dolomite. All screening was performed with distilled water to prevent any contamination and fouling of the resin, such as calcium sulfate being precipitated in the pores.

A selected size fraction was then preconditioned in a fluidized bed setup by osmotically cycling the resin material between the acid and salt form in order to flush out any residual impurities (solvents and functionalizing agents) and to stretch the resin matrix. For the cation exchange resin 2 N solutions of H^+ (acid form) and Na^+ (salt form) ions were used, while for the anion exchange resin 2 N solutions of OH^- (acid form) and Cl^- (salt form) ions were used. The conditioned size fraction was then rinsed with distilled water to remove any excess electrolyte solution before it was converted to the desired ionic form by exhaustive elution. A 2 N strength solution of the desired ion was used to convert the resin to a specific ionic form. Approximately 10 to 15 bed volumes of this desired 2 N solution were passed through the fluidized bed at a flow rate which resulted in a 10–15% bed expansion.

For experiments on the dolomitic material, a solution of hydrochloric acid was used instead of sulfuric acid in order to prevent precipitate formation in the pores of the resin. For the series of experiments on lead sulfate, the cation exchange resin was prepared in five different initial ionic forms (H^+, Na^+, K^+, Cu^{2+}, and Al^{3+}) and the anion exchange resin in three different ionic forms (Cl^-, NO_3^-, OH^-). The fully converted resin was subsequently continuously washed with distilled water in a fluidized bed setup until all the excess electrolyte solution was removed from the resin bead pores. This final rinsing step required 40 to 50 bed volumes of distilled water, being washed through the column at a fluidization flow rate that allowed for a 10% bed expansion. The resin was then stored in distilled water until used in the experiments.

C. Equilibrium Experiments

The equilibrium experiments were performed in 1 L high density PVC bottles. Separate experiments indicated significant contamination of the solution phase with elements such as Ca, Mg, Na, and K when glassware was used because glass often contains these elements. The low K_{sp} value of the dolomite ore resulted in very low aqueous phase concentrations of Ca and Mg ions (<1 ppm). Therefore, any Ca or Mg that leached from the glass into the solution would easily have resulted in a hundredfold increase in the concentration values reported. It was therefore important to avoid any contamination of any source to obtain reproducible results.

The bottles containing 1 L of distilled water, the resins, and the model solids were placed on rollers in a temperature-controlled water bath and agitated until complete equilibrium was obtained. Equilibrium curves were constructed by adding different measured volumes of either a cation or an anion exchange resin, in a specific ionic form, to the slurry containing the dolomite or lead sulfate. For the dolomite 90% of equilibrium was attained within 24 hours, with the final equilibrium being attained after one week. In the case of lead sulfate 1.365 g of $PbSO_4$ was used and equilibrium was attained after one month. The temperature was kept constant at 20°C during all experiments.

D. Analytical Methods

When complete equilibrium was attained, the resin-in-pulp mixture was separated into a clear filtrate solution and resin beads which were washed with distilled water to remove any excess mineral particles. The resin samples were washed, dried, crushed, weighed, ashed, dissolved by HNO_3, HF, and $HClO_4$ in a platinum crucible, and made up to a specific volume in a volumetric flask. The filtrate and dissolved resin sample solutions were then analyzed for the different cations using both atomic absorption and inductively coupled plasma spectrophotometry. The filtrate was also analyzed for the different anions using a Dionex ion chromatograph.

VII. EXPERIMENTAL CASE STUDY: DOLOMITE

For equilibrium solution concentrations (C_e) of calcium and magnesium in the order of 10^{-5} to 10^{-4} eq mol/L of solution, equilibrium loadings (q_e) of between 0.35 and 1.06 eq mol/L of free wet settled volume resin were obtained in the case of the gel-type strong acid cation exchange resin, Duolite C20. Equilibrium loadings of up to 0.7 eq mol/L fwsv were obtained on this resin for each of the calcium and magnesium species in its recovery from the dolomite ore, resulting in a total resin loading of 1.4 eq mol/L fwsv in equilibrium with solution concentrations of 10^{-4} eq mol/L for each of the species. These loadings were obtained with a single batch run that yielded recoveries of between 50 and 60% for each of the species. The strong acid macroporous-type cation exchange resin, Duolite C26, gave similar resin/solution and ore/solution behavior for the recovery of calcium and magnesium from a natural dolomite ore.

A. Dissolution Reactions with Only a Cation Exchange Resin

A small quantity of sodium (266 ppm in the ore) is present in the dolomite and will leach into the solution, but its concentration will be much lower than that of the calcium or magnesium. The competition between the leached sodium ions and either the calcium or the magnesium for sorption onto the ion exchange resin will be negligible. Thus the main competitive sorption will be between the calcium and magnesium ions leaching as follows:

$$CaCO_3 \overset{K_{sp}}{\leftrightarrow} Ca^{2+} + CO_3^{2-} \tag{17}$$

with $K_{sp} = 9.29 \times 10^{-9}$ for calcite at 20°C [15,16], and

$$MgCO_3 \overset{K_{sp}}{\leftrightarrow} Mg^{2+} + CO_3^{2-} \tag{18}$$

with $K_{sp} = 5.87 \times 10^{-5}$ for $MgCO_3$ at 20°C [15,16].

The solubility products K_{sp} for these two reactions were obtained from the equilibrium concentrations of calcium and magnesium in solution from dolomite leached in water with no ion exchange resin present. Figure 1 shows that the K_{sp} of calcium decreased and the K_{sp} of magnesium increased with increasing solid content of the slurry. The average values for the solubility products of calcium and magnesium were 3.42×10^{-8} and 4.12×10^{-8}, respectively. The solubility of natural dolomite (CaCO$_3$·MgCO$_3$; $M_r = 184.41$ g/mol) is given as 0.032 g per 100 ml water at 18°C [8], which gives a solubility product of 3.01×10^{-6} for each of the species.

The equilibrium concentrations (C_e) of calcium and magnesium in the solution phase were not constant [1], which is opposite to the expected trend if only one of the species were present in the solution at equilibrium

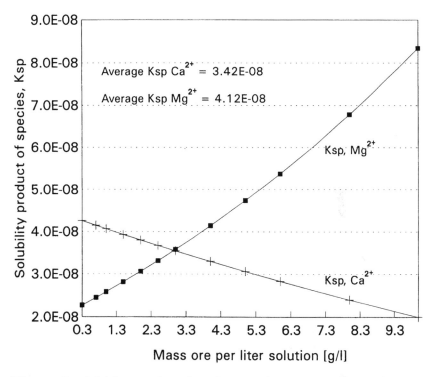

Figure 1 Solubility products for calcium and magnesium leached from a dolomitic ore.

(with its pure solid of low solubility). However, the sum of the concentrations of the two species was basically constant [1]. It therefore appears that the solubility product of the dolomite is constant, i.e., the sum total in equivalent moles of different species (calcium and magnesium) leaching is constant. The equilibrium concentration of sodium in solution was also constant [1].

Further reactions of importance are the ion exchange reactions where calcium and magnesium are exchanged for the counterion present on the cation exchange resin. The counterion initially on the resin was either Na^+ or H^+. These reactions are represented by Eq. (7). The values of the thermodynamic equilibrium constants as defined by Eq. (8) were determined experimentally for the Duolite C26 resin using completely soluble nitrate salts of calcium and magnesium, with the results being similar to those in the literature [17]:

$$K_{Na}^{Ca} = 2.5 \qquad K_{H}^{Ca} = 3.75$$

$$K_{Na}^{Mg} = 1.6 \qquad K_{H}^{Mg} = 2.45$$

Clearly the cation exchange resin was more selective toward calcium and magnesium than to sodium and hydrogen. Thus, as long as there are detectable traces of Ca^{2+} and Mg^{2+} ions in the solution, the exchange for sodium or hydrogen ions (initially on the ion exchanger) will continue until equilibrium is attained. The equilibrium constants for the exchange of either calcium or magnesium for the hydrogen counterion are larger than for the exchange of the sodium counterion. On this basis alone one expects higher equilibrium loadings of calcium and magnesium on the cation exchange resins initially saturated with H^+ ions than with Na^+ ions.

Most metal salts are soluble in water because of water's high dielectric constant and the ability of water to coordinate with the ions (solvate), especially with metal ions, so that the hydrated ions of the soluble salt more closely resemble the solvent [18]. The exchanged H^+ ions will be able to "merge" with the water (to form the "new" water solvent) to a larger extent than the Na^+ ions to approximately resemble the solvent (water) first because of their smaller ionic radii for the same valence, which results in a higher charged ion, and second due to free H^+ ions that are already a "building block" of water. Thus, more H^+ ions can merge into the water than Na^+ ions before affecting the hydration capability of the new water solvent with respect to the leached carbonate anions from the dolomite ore. The H^+ ion content of the solution can therefore be much higher than the Na^+ ion content for a given number of carbonate anions in solution. The higher the H^+ ion content (or Na^+ ion content) in solution, the more of these ions (H^+, Na^+) that initially saturate the ion exchange resin can be exchanged for either Ca^{2+} or Mg^{2+} ions. This leads to higher resin loadings of the valuable species.

For each Ca^{2+} or Mg^{2+} ion that is leached into solution, a CO_3^{2-} ion is released from the ore into solution that is solvated by the new solvent, which contains the exchanged counterion initially in the resin phase. It is therefore theoretically possible to obtain higher resin loadings with cation exchangers initially saturated with H^+ ions.

B. Enhanced Dissolution via Formation of a Weak Acid

If the ion exchange resin is initially saturated with H^+ ions, a weak acid forms between the H^+ counterion and the leached carbonate anions from the ore. The weak acid formed is carbonic acid (H_2CO_3), and the reaction is given by the following two equations:

$$H^+ + HCO_3^- \overset{K_1}{\leftrightarrow} H_2CO_3 \qquad \text{with } K_1 = 3.5 \times 10^{-7} \qquad (19)$$

$$H^+ + CO_3^{2-} \overset{K_2}{\leftrightarrow} HCO_3^- \qquad \text{with } K_2 = 6.0 \times 10^{-11} \qquad (20)$$

The second dissociation constant, K_2, of carbonic acid is so small that if any hydrogen ions are present in a solution with free carbonate anions, it will almost completely combine to form HCO_3^- according to Eq. (20). As the leached carbonate anions are protonated with the exchanged hydrogen ions from the ion exchanger, more of the solid must dissolve, according to Le Chatelier's principle, to restore the free carbonate anion content of the solution. The higher the exchanged hydrogen ion concentration in the solution, the greater the solubility of the solid becomes, as explained earlier with Eq. (16). Therefore, it is possible to obtain higher resin loadings of calcium and magnesium on a cation exchange resin initially saturated with H^+ ions, as counterions, than one saturated with Na^+ ions.

The effect of the cation resin state is depicted in Fig. 2, which shows the maximum amount of calcium and magnesium extracted for a specific mass of dolomite ore in contact with 1 L of solution. It is evident that for both the extraction of calcium and magnesium, the macroporous cation exchange resin initially saturated with H^+ ions resulted in a higher equilibrium extraction value than when saturated with Na^+ ions.

If the resin is initially saturated with sodium ions, it is evident from the following reactions and equilibrium constants that free sodium ions

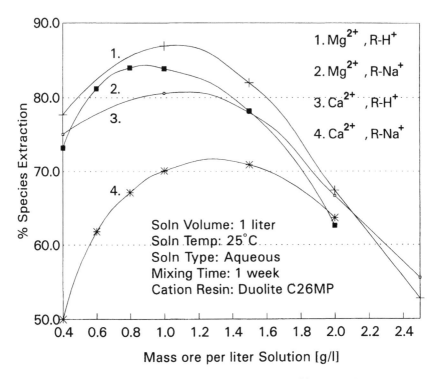

Figure 2 The effect of the cation resin state on equilibrium resin recoveries of calcium and magnesium from a dolomitic ore using Duolite C26. (From Ref. 1. Copyright 1995, with permission from Elsevier Science.)

will combine with free carbonate ions only at very high concentrations (7000 to 9000 times higher than that needed with H^+ ions).

$$Na_2CO_3 \cdot H_2O \overset{K}{\leftrightarrow} 2Na^+ + CO_3^{2-} + H_2O \qquad (21)$$

with $K = 2.70 \times 10^2$ at 30°C [15,16];

$$NaHCO_3 \overset{K}{\leftrightarrow} Na^+ + HCO_3^- \qquad (22)$$

with $K = 1.31 \times 10^0$ at 20°C[15,16].

Tables 2 and 3 show equilibrium data for the extraction of calcium and magnesium from a dolomite ore using a macroporous and gel-type cation exchange resin with different mixed bed configurations using dif-

Table 2 Equilibrium Data for the Extraction of Calcium and Magnesium from a Dolomitic Ore with a Macroporous Cation Exchanger (Duolite C26) in a Mixed Bed with Different Volumes of a Macroporous Anion Exchanger (A161)

Volume SBA[a] (mL)	Mass ore (g/L)	Equilibrium loading (eq mol/L fwsv)			Equilibrium concentration (eq mol/L solution)			Extraction (%)	
		Ca	Mg	Na	Ca	Mg	Na	Ca	Mg
A161		*Strong base anion resin*			*Resin state: $R^+ Cl^-$*			*$dp = 1180$ μm*	
20 mL C26		*Strong acid cation resin*			*Resin state: $R^- Na^+$*			*$dp = 922$ μm*	
5	1.001	0.14	0.10	1.26	2.39E-6	6.00E-6	4.61E-3	26.15	17.91
10	1.002	0.16	0.11	1.23	4.39E-6	7.90E-6	2.78E-3	30.43	20.81
20	1.004	0.19	0.13	1.24	2.10E-5	1.14E-5	1.62E-3	32.58	22.39
A161		*Strong base anion resin*			*Resin state: $R^+ Cl^-$*			*$dp = 1180$ μm*	
20 mL C26		*Strong acid cation resin*			*Resin state: $R^- H^+$*			*$dp = 922$ μm*	
5	1.00	0.45	0.45	1.48	5.50E-7	3.95E-6	6.52E-7	84.18	83.47
10	1.00	0.44	0.45	1.12	1.99E-7	2.39E-6	3.91E-7	81.86	82.59
20	1.00	0.46	0.50	0.96	2.50E-7	1.81E-6	2.83E-6	85.07	91.91
A161		*Strong base anion resin*			*Resin state: $R^+ OH^-$*			*$dp = 1180$ μm*	
20 mL C26		*Strong acid cation resin*			*Resin state: $R^- Na^+$*			*$dp = 922$ μm*	
5	1.00	0.14	0.10	1.47	1.64E-6	5.36E-6	4.41E-3	27.14	18.19
10	1.00	0.19	0.12	1.42	2.58E-6	6.41E-6	2.17E-3	33.34	22.91
20	1.00	0.23	0.17	1.34	1.21E-5	2.14E-5	1.25E-3	42.65	31.23
A161		*Strong base anion resin*			*Resin state: $R^+ OH^-$*			*$dp = 1180$ μm*	
20 mL C26		*Strong acid cation resin*			*Resin state: $R^- H^+$*			*$dp = 922$ μm*	
5	1.005	0.44	0.46	2.71	2.25E-7	2.06E-6	1.96E-7	80.95	84.09
10	1.001	0.39	0.41	2.85	9.74E-6	3.70E-6	6.29E-7	71.13	75.06
20	1.002	0.24	0.25	4.89	13.2E-7	15.8E-6	2.17E-7	43.24	45.74

[a] SBA: strong base anionic resin.
Note. Both exchangers are used in different initial resin states.
Source. Ref. 1.

Table 3 Equilibrium Data for the Extraction of Calcium and Magnesium from a Dolomitic Ore with a Gel-Type Cation Exchanger (Duolite C20) in a Mixed Bed with Different Volumes of a Macroporous Anion Exchanger (A161)

Volume SBA[a] (mL)	Mass ore (g/L)	Equilibrium loading (eq mol/L fwsv)			Equilibrium concentration (eq mol/L solution)			Extraction (%)	
		Ca	Mg	Na+	Ca	Mg	Na	Ca	Mg
A161 *20 mL C20*		*Strong base anion resin* *Strong acid cation resin*			*Resin state: $R^+ Cl^-$* *Resin state: $R^- Na^+$*			$dp = 1180\ \mu m$ $dp = 922\ \mu m$	
5	1.002	0.15	0.10	1.40	1.10E-6	2.72E-6	4.01E-3	28.19	18.10
10	1.004	0.17	0.11	1.41	1.25E-6	2.96E-6	4.70E-3	32.24	20.35
20	1.001	0.19	0.12	1.51	1.85E-6	5.43E-6	5.52E-3	34.85	21.94
A161 *20 mL C20*		*Strong base anion resin* *Strong acid cation resin*			*Resin state: $R^+ Cl^-$* *Resin state: $R^- H^+$*			$dp = 1180\ \mu m$ $dp = 922\ \mu m$	
5	1.000	0.44	0.47	9.78	2.99E-7	2.63E-6	1.35E-6	81.65	85.86
10	1.008	0.46	0.48	4.88	4.88E-7	3.79E-6	7.35E-6	83.76	87.24
20	1.005	0.44	0.48	4.26	4.99E-8	1.56E-6	1.17E-6	81.74	87.13
A161 *20 mL C20*		*Strong base anion resin* *Strong acid cation resin*			*Resin state: $R^+ OH^-$* *Resin state: $R^- Na^+$*			$dp = 1180\ \mu m$ $dp = 922\ \mu m$	
5	1.009	0.15	0.10	1.46	9.98E-7	2.14E-6	3.88E-3	28.22	18.91
10	1.001	0.19	0.13	1.43	1.40E-6	3.29E-6	5.46E-3	34.65	23.64
20	1.001	0.24	0.16	1.35	2.35E-6	5.10E-6	7.15E-3	45.21	29.05
A161 *20 mL C20*		*Strong base anion resin* *Strong acid cation resin*			*Resin state: $R^+ OH^-$* *Resin state: $R^- H^+$*			$dp = 1180\ \mu m$ $dp = 922\ \mu m$	
5	1.004	0.43	0.46	10.5	9.98E-7	1.05E-5	1.35E-6	80.11	85.09
10	1.005	0.33	0.36	9.90	9.48E-7	8.72E-6	2.31E-6	61.70	65.59
20	1.002	0.21	0.24	9.41	3.49E-7	4.77E-6	4.78E-7	39.57	44.32

[a] SBA: Strong base anionic resin.

Note: Both exchangers are used in different initial resin states.

Source: Ref. 1.

ferent volumes (5, 10, 20 mL) of a macroporous anion exchange resin (Duolite A161). Both the cation and anion exchange resins were used in different initial resin states and were added to 1 L of water in contact with 1 g of the dolomite. From Tables 2 and 3 it is evident that higher extractions were obtained for both calcium and magnesium in a mixed bed of ion exchangers when the cation exchange resin was initially saturated with H^+ ions (R^-H) instead of Na^+ ions (R^-Na^+). Consequently, in both a mixed bed configuration and a cation resin bed only, higher equilibrium extractions were possible when the cation exchange resin was initially saturated with H^+ ions.

The solubility of the solid could therefore be enhanced if the co-ion initially saturating the ion exchange resin (cation or anion resin in case of mixed bed) forms a weak acid (or base) with the undesirable ions that also leach from the ore. This results in an additional driving force to enhance the overall dissolution of the solid, coupled with the sorption of the leached valuable species.

C. Gel-Type Versus Macroporous Cation Exchange Resins

Usually a gel-type resin is capable of higher resin loadings (in equilibrium with high solution concentrations) and hence could extract more of the valuable species for the same resin volume used at high solution concentrations than in the case of the macroporous-type resin. However, the low solubility of the dolomite resulted in very low solution concentrations of the leached species, so that the resin loadings were seldom near the saturated loadings. Consequently, both the gel- and macroporous-type cation exchange resins yielded similar loadings and percentage extraction of the calcium and magnesium, as seen in Table 4.

Figure 3 shows that the differences in the equilibrium resin loading and extraction percentages obtained for both calcium and magnesium between a gel- and macroporous-type cation exchanger are negligible for the concentration region of interest. The extraction curves show a maximum value for each of the species using either a gel- or macroporous-type exchange resin. Tables 2 and 3 show equilibrium data for the extraction of calcium and magnesium from a dolomite ore using either a macroporous- or gel-type cation exchange resin. Comparing the values for the percentage

Table 4 Equilibrium Data for the Extraction of Calcium and Magnesium from a Dolomitic Ore with a Gel-(Duolite C20) and Macroporous Type (Duolite C26) Cation Exchange Resin for Slurries with Different Contents of Solids

Mass ore (g/L)	Equilibrium loading (eq mol/L fwsv)			Equilibrium concentration (eq mol/L solution)			Extraction (%)	
	Ca^{2+}	Mg^{2+}	Na^+ ($\times 10^3$)	Ca^{2+}	Mg^{2+}	Na^+	Ca^{2+}	Mg^{2+}
Strong acid cation resin: 20 mL C26				*Resin state: $R^- H^+$*			*dp = 922 μm*	
2.000	0.686	0.664	1.810	6.62E-4	2.24E-3	4.45E-5	63.65	61.10
1.001	0.435	0.473	4.223	6.99E-7	9.47E-6	2.18E-7	80.68	86.95
0.410	0.166	0.174	7.393	1.09E-6	6.35E-6	1.16E-6	75.06	77.71
Strong acid cation resin: 20 mL C20				*Resin state: $R^- H^+$*			*dp = 922 μm*	
2.502	0.753	0.726	2.908	9.76E-4	2.59E-3	9.22E-4	55.84	53.48
1.004	0.441	0.479	8.949	6.49E-7	6.99E-6	5.65E-7	81.54	87.86
0.402	0.151	0.150	6.153	1.50E-7	1.48E-6	1.48E-6	69.81	68.75

Source: Ref. 1.

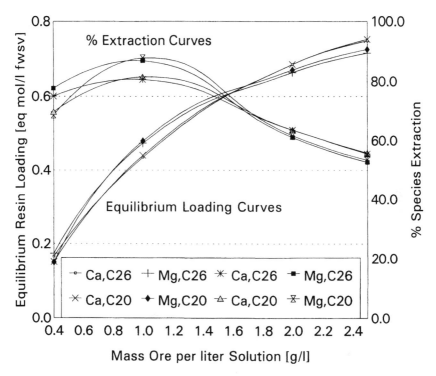

Figure 3 Equilibrium extraction and resin loading curves of calcium and magnesium from a dolomitic ore with a macroporous- (Duolite C26) and gel-type (Duolite C20) cation exchanger. (From Ref. 1. Copyright 1995, with permission from Elsevier Science.)

calcium and magnesium extracted for the same mixed bed configurations, it is evident that no significant difference exists between the equilibrium extraction values of a gel-type and macroporous-type cation exchange resin for this application.

D. Effect of the Anion Resin State in a Mixed Configuration

The effect of the counterion of the anion exchange resin in a mixed bed is dependent on the resin state of the accompanying cation exchange resin. This effect can be seen in Tables 2 and 3, which show the extraction of calcium and magnesium for different volumes of the anion exchanger used in the mixed bed. If the cation counterion is the Na$^+$ ion, the OH$^-$ ion is the better counterion for the anion exchanger than the Cl$^-$ ion. If the H$^+$ ion is the cation counter ion, the Cl$^-$ ion is the better anion counterion thus resulting in higher extraction percentages of calcium and magnesium than with the OH$^-$ ion.

Figure 4 shows that a significant reduction was observed in the extraction of magnesium when the anion exchanger in the mixed bed was used in the free base form. Similar behavior was observed for the calcium [1]. The anion exchanger in the free base form will remove any of the free anion species available in the solution without replacing the removed species with any counter anions. This removal of anions will create a solu-

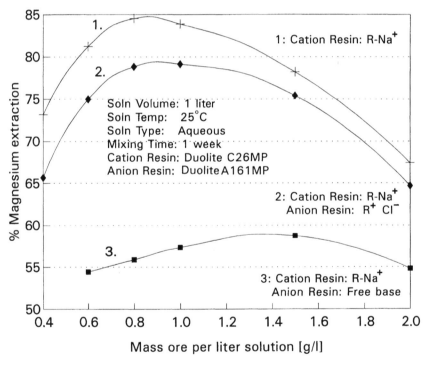

Figure 4 Equilibrium extraction curves for magnesium from a dolomitic ore with different ion exchange mixed bed configurations of Duolite C26 and Duolite A161.

tion with a "net positive" charge. Cationic exchange proceeds on the basis of equivalent charged moles, meaning that the net positive charge of the solution cannot be removed by the cation exchange reaction. This net positive charged solution will prevent the further leaching of any more valuable cations from the ore. Electrical neutrality in any natural solution is always maintained, and in this case it leads to the precipitation of the leached cations, resulting in the lower recoveries.

VIII. EXPERIMENTAL CASE STUDY: LEAD SULFATE

Equilibrium resin loading isotherms for the sorption of Pb^{2+} ions from aqueous solutions by a cation exchange resin, Duolite C26, used in different initial ionic forms are depicted in Fig. 5. The ionic forms included the H^+,

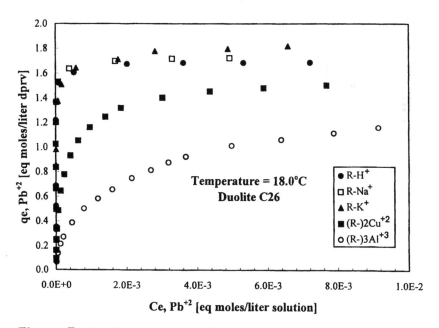

Figure 5 Equilibrium isotherms for the sorption of Pb^{2+} ions by a strong acid cation exchange resin (Duolite C26) from clear solutions of $Pb(NO_3)_2$ for different initial cationic forms of the resin at 18.0°C. (From Ref. 2. Copyright 1997, with permission from Elsevier Science.)

Na$^+$, K$^+$, Cu^{2+}, and Al^{2+} cations. The rectangularly shaped isotherms for the H$^+$, Na$^+$, and K$^+$ are typical of modern synthetic ion exchangers, which clearly indicates that a substantially high resin loading of the valuable species can be obtained for very low solution phase concentrations of the valuable species. Duolite C26 is a relatively nonselective strong acid cation exchange resin with sulfonic functional groups. In the case of resins specifically designed to be selective toward a particular species, these equilibrium isotherms can become much more rectangular in shape for the sorption of the specific ion. With resins of this type, near maximum loadings can be obtained for the desired species with much smaller total solution phase concentrations (including other competing species).

It is well documented that the selectivity of a resin for a particular species increases with an increase in the charge density of that species. This is evident from Fig. 5, which shows a significant decrease in the sorption of the Pb^{2+} ions with an increase in the valence of the counterion initially saturating the cation exchange resin, especially at the lower solution phase concentrations. This phenomenon can be used to selectively extract valuable species from a multicomponent ore if a significant difference exists in the charge density of the various species.

For each mole of cationic species B dissolved from the solid, an equivalent number of moles of anionic species Y accompanies it, as in Eq. (1), in order to maintain electrical neutrality of the solution phase. The anions of species Y are largely excluded from the cationic resin phase and are not removed from the solution phase. The consequence is a steady increase in its solution phase concentration as the ion exchange reaction proceeds, which leads to a decrease in the solubility of the sparingly soluble solid due to the common ion effect. This is evident from Fig. 6, which shows a decrease in the solution phase concentration of the Pb^{2+} ions (species B) in a PbSO$_4$ slurry containing no resin, as the solution phase concentration of the SO$_4^{2-}$ ions (species Y) is increased with the addition of Na$_2$SO$_4$.

The same phenomenon is shown by Fig. 7, caused by the ion exchange reaction which resulted in the formation of different amounts of H$_2$SO$_4$, Na$_2$SO$_4$, K$_2$SO$_4$, CuSO$_4$, and Al$_2$(SO$_4$)$_3$ for the different counterions initially saturating the resin. Figure 7 gives the equilibrium solution phase concentrations of the Pb^{2+} and SO$_4^{2-}$ ions in an aqueous slurry of PbSO$_4$ by changing the amount of the cation exchange resin, Duolite C26. As expected, the Al^{3+} and Cu^{2+} forms show higher concentrations of Pb^{2+}. Figure 8 shows the corresponding equilibrium solution phase

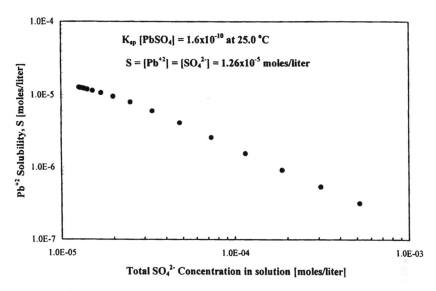

Figure 6 Decrease in the equilibrium solution concentration of Pb^{2+} ions in a slurry of $PbSO_4$ with an increase in the SO_4^{2-} concentration, showing the common ion effect. (From Ref. 2. Copyright 1997, with permission from Elsevier Science.)

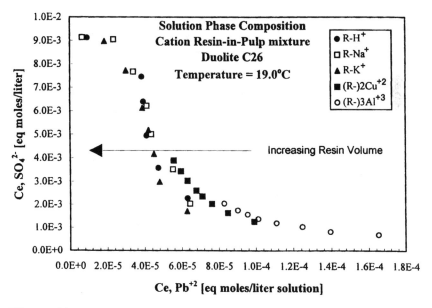

Figure 7 Equilibrium solution phase concentrations of Pb^{2+} and SO_4^{2-} ions in an aqueous $PbSO_4$ RIL slurry containing a strong acid cation exchange resin (Duolite C26) for different initial cationic forms of the resin at 19.0°C. (From Ref. 2. Copyright 1997, with permission from Elsevier Science.)

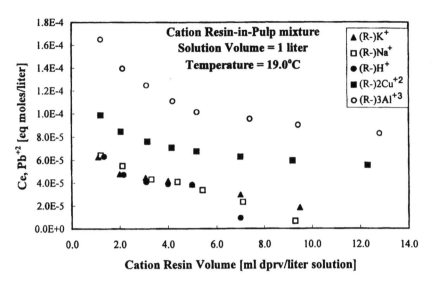

Figure 8 Equilibrium solution phase concentrations of Pb^{2+} ions in an aqueous $PbSO_4$ slurry mixture as a function of the volume of cationic resin (Duolite C26) added for different initial cationic forms of the resin at 19.0°C. (From Ref. 2. Copyright 1997, with permission from Elsevier Science.)

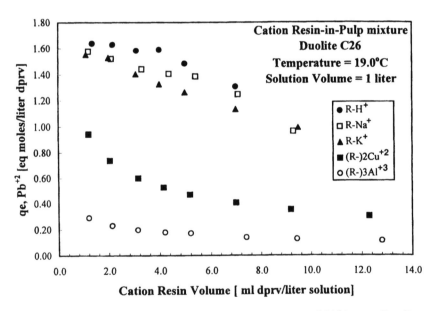

Figure 9 Equilibrium resin loading for the sorption of Pb^{2+} onto Duolite C26 from an aqueous slurry of $PbSO_4$ as a function of the volume of resin added for different initial cationic forms of the resin at 19.0°C. (From Ref. 2. Copyright 1997, with permission from Elsevier Science.)

concentrations of the Pb^{2+} ions for the different volumes of the cationic resin used. Figure 9 shows the equilibrium loadings of the Pb^{2+} ions on the resin for the different volumes of the ion exchanger used. When comparing Figs. 8 and 9, it is evident that the highest solution phase concentrations of the Pb^{2+} ions were obtained for the Al^{3+} form of the cationic resin, which corresponded to the lowest equilibrium loading obtained for the Pb^{2+} ions across the whole range of cationic resin volumes used. This trend is confirmed by Figs. 5 and 10, which show lower equilibrium loadings of the Pb^{2+} ions for an increase in the valence of the counterion. The solution phase concentrations of the Pb^{2+} ions in the slurry mixtures in Fig. 10 correspond to the lower concentration region of Fig. 5, where the shape of the isotherm is more linear.

Figure 8 shows that higher solution phase concentrations of the Pb^{2+} ions were obtained with the cationic resin used in the higher valence state.

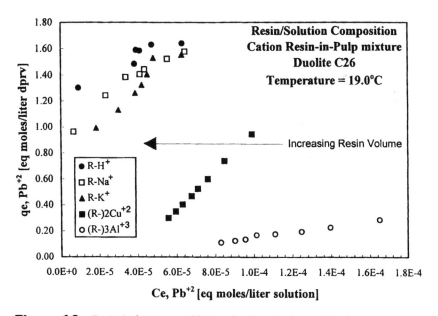

Figure 10 Resin/solution equilibrium loading isotherms for the sorption of Pb^{2+} ions onto Duolite C26 from a slurry of $PbSO_4$ for different initial ionic forms of the cation exchange resin at 19.0°C. (From Ref. 2. Copyright 1997, with permission from Elsevier Science.)

The reason for this is that the exchanged counterion of higher charge density ($Al^{3+} > Cu^{2+} > H^+ > K^+ > Na^+$) affects the solubility of the sparingly soluble solid more strongly. As a rule of thumb, stronger ion–ion interactions are observed between ions of higher charge density because of the larger electrochemical potential difference between the mentioned cations and the contaminant species (SO_4^{2-} ions). These interactions cause the solution to become more nonideal, which is represented by a deviation in the activity coefficients of the species from unity. According to Eq. (7), this results in an increase in the solubility of the sparingly soluble solid as the dissolved SO_4^{2-} ions are "masked from being recognized" by the Pb^{2+} ions in the dissolution reaction due to "cluster forming" with the oppositely charged counterion A. This phenomenon was mentioned earlier, when it was stated that the solubility of the sparingly soluble solid can be affected by the formation of new electrolyte solutions as a result of the ion exchange reaction. However, Fig. 10 clearly indicates that much higher resin loadings of the Pb^{2+} ions are obtained with the resin in the monovalent ionic form compared to the bivalent and trivalent forms, outweighing the increase in loading obtained from enhancement of electrolyte solubility of the bivalent and trivalent ionic forms.

A higher resin loading can be obtained if the amount of the resin added to a batch reactor is decreased, which will decrease the common ion effect, as illustrated by Fig. 9. In order to minimize the common ion effect, a mixture of cationic and anion exchange resins can be used to simultaneously remove both the cations and anions, which prevents a significant increase in the concentration of the constituent ions of the sparingly soluble solid. Figures 11 and 12 show the attainable equilibrium loading of Pb^{2+} ions onto the cationic resin, Duolite C26, when a mixture of a cationic and anionic resin is used in the slurry of $PbSO_4$. In Fig. 11 the cationic resin was initially in the K^+ ionic form and the anion resin, Duolite Al61, in the NO_3^- ionic form. In Fig. 12 the Al^{3+} and NO_3^- forms of the cationic and anionic resin were used. It is evident from both Figs. 11 and 12 that the cationic resin loading (Pb^{2+} ions) increases, while the anionic resin loading (SO_4^{2-} ions) decreases with an increase in the ratio of the volume of the anionic resin used relative to the volume of the cationic resin used. Different curves were constructed for different total volumes of the cation exchange resin.

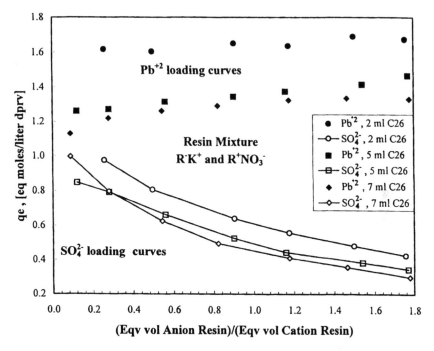

Figure 11 Equilibrium resin loading of Pb^{2+} ions onto Duolite C26 from an aqueous slurry of $PbSO_4$ containing the cationic resin in the K^+ ionic form and the anion resin, Duolite A161, in the NO_3^- ionic form, as a function of the ratio of the equivalent volume of anionic resin to equivalent volume of cationic resin. (From Ref. 2. Copyright 1997, with permission from Elsevier Science.)

Another important feature of the simultaneous dissolution and sorption RIL process is shown by Fig. 13, which indicates that the overall ionic concentration of the solution increases with an increase in the volume of the ion exchange resin used. Figure 13 shows the solution phase concentrations of the Pb^{2+}, K^+, and SO_4^{2-} ions as well as the total solution phase molar ionic strength, I, defined in Eq. (23):

$$I = \frac{1}{2} \sum_i |Z_i|^2 C_i \qquad (23)$$

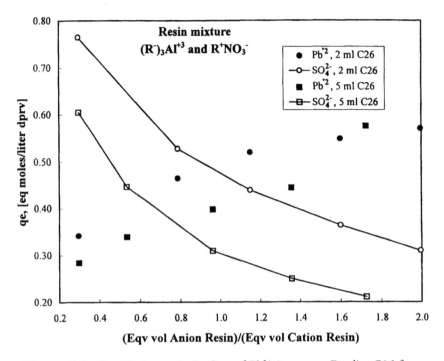

Figure 12 Equilibrium resin loading of Pb^{2+} ions onto Duolite C26 from an aqueous slurry of $PbSO_4$ containing the cationic resin in the Al^{3+} ionic form and the anion resin, Duolite A161, in the NO_3^- ionic form, as a function of the ratio of the equivalent volume of anionic resin to equivalent volume of cationic resin. (From Ref. 2. Copyright 1997, with permission from Elsevier Science.)

This increase in the solution phase ionic concentration of the counter cation affects the activity of the slightly dissolved anions of the solid, as mentioned earlier. This results in a change in the activity coefficient of the anion, which may lead to either an increase or a decrease in the solubility of the sparingly soluble solid. Figure 14 shows the total solution molar ionic strength, I, for the different initial ionic forms of the cation exchange resin for increasing amounts of resin. It is evident that a steeper increase exists for the monovalent ionic forms as compared to the bivalent and trivalent forms, which is caused by the higher solution phase concentrations of the monovalent counterions. This increase in the total molar ionic strength of the solution phase with an increase in the total amount of

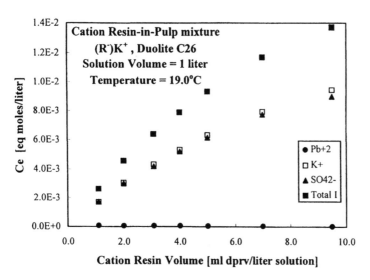

Figure 13 Total molar solution ionic strength and equilibrium solution phase concentrations of Pb^{2+}, SO_4^{2-}, and K^+ in an aqueous slurry of $PbSO_4$ containing Duolite C26 initially in the potassium ionic form $(R^-)K^+$, as a function of the volume of resin added at 19.0°C. (From Ref. 2. Copyright 1997, with permission from Elsevier Science.)

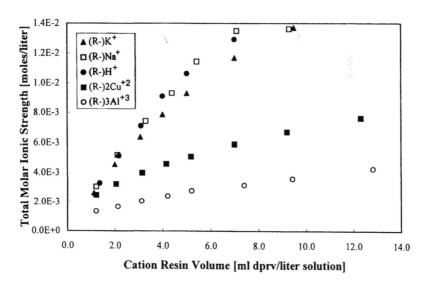

Figure 14 Total molar solution ionic strength in an aqueous slurry of $PbSO_4$ containing Duolite C26 in different initial ionic forms as a function of the volume of resin added at 19.0°C. (From Ref. 2. Copyright 1997, with permission from Elsevier Science.)

resin added can be used to create concentrated electrolyte solutions. Such electrolyte solutions can be controlled to contain relatively small amounts of the constituent ions of the sparingly soluble solid in a RIL mixture incorporating both a cation and anion exchange resin by carefully manipulating the ratio of the resins. The solubility of the sparingly soluble solid can be enhanced significantly in this way, which illustrates the flexibility of the RIL process.

IX. CONCLUSIONS

The development of highly selective ion exchange resins of high capacity is causing a paradigm shift in the recovery of valuable species from dissolving solids. Consequently, the simultaneous dissolution and sorption of species from sparingly soluble solids in resin-in-leach is a promising hydrometallurgical process. Considerable savings in reagent consumption can be achieved by concentrating the valuable species in a particulate extractive phase without the use of a leaching reagent.

Ion exchange resins are also capable of significantly enhancing the leaching of valuable species from sparingly soluble mineral particles and waste by continually removing the leached valuable species from the solution, hence disturbing the solid/liquid equilibrium condition, which in turn will be re-established by further dissolution of the valuable species. This process is feasible as a result of the ability of modern ion exchange resins to attain high loadings in equilibrium with extremely low solution phase concentrations, as encompassed in the rectangular shape of a typical ion exchange isotherm.

Improved extraction of the valuable mineral can be achieved if the associated contaminant species are removed from the solution phase during the leaching procedure. In conventional leaching circuits, this improved efficiency is achieved by dividing the leaching procedure into different stages, inevitably leading to higher operational costs. In the case of the RIL system the contaminant concentration in the leaching stage can be reduced with the correct choice of resin type and initial counterion content of the resin phase. Precipitation and weak acid formation reactions between the contaminant and the exchanged counterion can be used to reduce the concentration of the contaminants. In addition, mixtures of cation and anion exchange resins can be used to reduce the concentration of contaminants in solution. The overall recovery of a valuable species can further be improved if its solubility can be enhanced. This can be achieved by creating electrolyte solutions from the exchanged counterions initially saturating the cation and anion exchange resins in a RIL mixture.

Unfortunately, this promising technology has not been commercialized widely. Further advances in the synthesis of ion exchange resins and inorganic adsorbents having highly selective functional groups will help to promote these simultaneous leaching and sorption processes. It is also possible that environmental restrictions on the use of leaching reagents will provide the final impetus for the acceptance of this new paradigm.

NOMENCLATURE

a_i	Solution phase activity of species i
$\bar{a}_i$	Resin phase activity of species i
A	Counterion initially saturating the cation exchange resin
B	Constituent cation of the precipitate
C_e	Equilibrium solution phase concentration
DPRV	Densely packed resin volume, i.e., the volume of resin that is compacted by tapping the measuring vessel until no more consolidation is observed during measurement
fwsv	Free wet settled volume of resin, i.e., the volume when resin is allowed to settle undisturbed and only under the force of gravity
K_{sp}	Solubility product constant
K_a	Dissociation constant of weak acid
K_A^B	Thermodynamic equilibrium constant for the cation exchange resin
K_{diss}	Thermodynamic equilibrium dissolution constant
m	Increase in the concentration of the common ion
R	Represents the solid resin matrix with the functional groups attached, denoted as R^-Cation or R^+Anion in the case of a cation and anion exchange resin, respectively
S	Solubility
Q	Resin capacity

q_e	Equilibrium loading on resin
y_i	Resin phase mole fraction of species i
Y	Constituent anion of the precipitate
z_i	Ionic valence of species i

Greek

ε	Volumetric fraction of interstitial water in a measured volume of resin
ν_i	Stoichiometric number of moles of species i formed when one mole of precipitate completely dissolves
γ_i	Solution phase activity coefficient of species i
$\bar{\gamma}_i$	Resin phase activity coefficient of species i

Subscripts

i	indicates species i
s	solid phase

REFERENCES

1. PGR De Villiers, JSJ van Deventer, L Lorenzen. Miner Eng 8(11):1309–1326, 1995.
2. PGR De Villiers, JSJ van Deventer, L Lorenzen. Miner Eng 10(9):929–945, 1997.
3. T Nishino. Dissolution Process of Portland cement powder in aqueous suspension containing ion exchange resin. New developments in ion exchange. Materials, fundamentals, and applications. Proceedings of the International Conference on Ion Exchange, ICIE '91, Tokyo, October 2–4, 1991. Tokyo: Kodansha, 1991, pp 3–6.
4. S Yang, H Gao, Y Su. Study on adsorption equilibrium of sorbic acid from dilute solution by TRPO extracting resin: New developments in ion exchange. Materials, fundamentals, and applications. Proceedings of the International Conference on Ion Exchange, ICIE '91, Tokyo, October 2–4, 1991. Tokyo: Kodansha, 1991, pp 31–34.
5. FLD Cloete, AP Marais. Ind Eng Chem Res 34:2464–2467, 1995.
6. V Kadlec, P Huber, CKD Dukla. Combined Ion Exchange—Precipitation Process for Minimizing Liquid Wastes in Water Treatment. Ion Exchange For Industry. Chichester, England: Ellis Horwood, 1988, pp 148–155.

7. S Brewer. Solving Problems in Analytical Chemistry. New York: Wiley, 1980, pp 218–254.
8. F Helfferich. Ion Exchange. New York: McGraw-Hill, 1962, pp 226–229.
9. AK Chakravarti, G Fritzsch. Reactive Polymers 8:51–68, 1988.
10. A de Lucas Martinez, P Canizares, J Zarca Diaz. Chemical Eng Technol 16:35–39, 1993.
11. AM Elprince, KL Babcock. Soil Science 120(5):332–338, 1975.
12. H Kim, WJ Frederick, Jr. J Chem Eng Data 33:177–184, 1988.
13. H Kim, WJ Frederick, Jr. J Chem Eng Data 33:278–283, 1988.
14. MA Mehablia, DC Shallcross, GW Stevens. In: MJ Slater, ed. Proceedings of IEX '92. London: Elsevier Applied Science, 1992, pp 151–158.
15. MJ Astle, RC Weast. CRC Handbook of Chemistry and Physics. 62nd ed. Boca Raton, FL: CRC Press, 1981–1982, pp D231–D232, B87–B88, B115–B117, B145–B151, B242–B243.
16. HP Perry, DW Green, JO Maloney. Perry's Chemical Engineer's Handbook. 6th ed. New York: McGraw-Hill, 1984, pp (3-97)–(3-100).
17. F de Dardel, TV Arden. Ion Exchange. Principles and Applications. Separation Technologies. Vol A14, 1989.
18. GD Christian, JE O'Reilly. Instrumental Analysis. 2nd ed. Needham Heights, MA: Allyn & Bacon, 1990, pp 642–648.

3

Polymeric Ligand-Based Functionalized Materials and Membranes for Ion Exchange

Stephen M. C. Ritchie and Dibakar Bhattacharyya

University of Kentucky, Lexington, Kentucky

I. INTRODUCTION

The separation and recovery of metals and other inorganics is a field of considerable interest. Many technologies exist to remove and recover metals/oxyanions from solution. These can be as varied as precipitation by chemical addition to separation with well-established membrane techniques, such as reverse osmosis and ion exchange [1] membranes. However, to selectively recover the entire spectrum of inorganics, the use of specific interactions with chemical groups has proven to be very effective, particularly in the low concentration range. These processes are most commonly referred to as ion exchange and chelation. Research in this area has been extensive and has been documented in classic publications [2] and in current reviews [3–5]. Interaction between researchers in academia, industry, and the government has also been intense, such as with the recent Metals Adsorption Workshop in Cincinnati, OH [6].

II. CONVENTIONAL ION EXCHANGE

Ion exchange often involves very specific interactions between the ion in solution and a group in the solid phase. The most common substrates for the interaction of ions and functional groups are ion exchange resins. This involves the construction of a matrix (most commonly polymeric) containing functional groups. Conventional monomeric functional groups for ion exchange include sulfonic acid, carboxylic acid, and amines. Groups may be acidic or basic, though in all cases, ions in solution are attracted to the substrate by electrostatic interactions. Depending on the charge and hydrated radius of the ion, the resulting complex will have a stability generally greater than that between the resident co-ion and exchange group.

Ions are recovered from the resin by driving this equilibrium reaction in the opposite direction. The resin is typically flooded with an excess of the resident co-ion such that the sorbed ion may diffuse from the resin and be recovered in the so-called regenerant solution. The concentration of recovered species in the regenerant solution is greater than its original concentration in the feed solution. Minimizing the volume of regenerant solution is a major concern in this field. It has also been one of the leading causes for research into macroreticular resins, where accessibility is enhanced. Hence, ions are more easily regenerated, and less regenerant solution may be utilized.

Ion exchange resins, as mentioned, are typically functionalized polymeric matrices. Functional groups, located on a typically microporous (<10 nm pore size) polymer matrix, interact with ions that diffuse into the matrix. Hence, several considerations must be made when utilizing ion exchange. Since ions must diffuse into the matrix, an adequate residence time is required. Optimum ion exchanger design should involve the use of pore characteristics such that high accessibility for interaction sites can be obtained with minimum pressure drop. The use of macroreticular resins has helped to overcome some of these problems.

A. Selectivity

Another area of importance when dealing with ion exchange is selectivity. Solutions are rarely idealized with only a few select noninterfering ions.

Therefore, interactions between species in solution and competitive sorption must be considered. In general, since ion exchange works by simple electrostatic interactions, little selectivity is inherent in the process. One way that researchers have incorporated selectivity, particularly in the case of quarternary amines, is to use steric hindrance to exclude bulky anions (Fig. 1). For example, a typical quarternary amine will always have a higher selectivity for sulfate over nitrate due to its larger negative charge. However, when the functional group is a tributyl substituted amine, sulfate cannot access the group easily, and nitrate can be selectively sorbed [7,8].

Chelation groups have also been utilized to increase the selectivity of ion sorbing resins. This is particularly useful for the sorption of transition metals, where solutions may contain hardness (i.e., Ca^{2+}, Mg^{2+}) that can compete for available sorption sites. It should be noted that chelating resins are used for cation sorption only. The mechanism of sorption occurs both by electrostatic interactions and by interaction of the sorbed cation with a lone pair of electrons on the functional group. In general, the lone pair of electrons is supplied by a nitrogen on the group, though phosphorus and sulfur may also "donate" the electrons. The result is a multidentate complex that is inherently more stable than ion exchange alone. Stabilities of some selected chelate–metal complexes are shown in Table 1. One example of a chelating group is iminodiacetic acid (IDA). The carboxylic acid groups in IDA provide the electrostatic interactions, while the nitrogen group of the amine provides an electronegative center that interacts with electron poor ions. It should be noted that chelates complex metal ions more strongly than ion exchange, as evidenced by sulfamic acid in Table 1. This ligand is composed of a primary amine group and a sulfonic

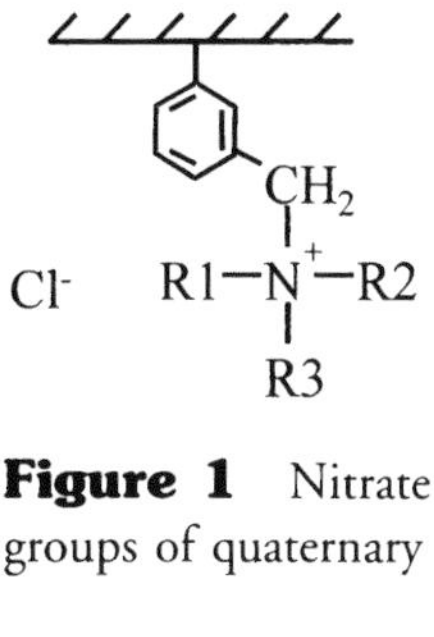

SO_4^{2-}/NO_3^- Selective: R1,R2,R3 = CH_3
Low NO_3^-/SO_4^{2-} Selective: R1,R2,R3 = C_2H_5
High NO_3^-/SO_4^{2-} Selective: R1,R2,R3 = C_4H_9

Figure 1 Nitrate selectivity incorporation by modification of substituted groups of quaternary amines. (From Ref. 7.)

Table 1 Metal Complex Stability Constants (log K for $[ML]/[M][L]$) for Various Chelating Agents

	Cu	Pb	Cd	Ca
Aspartic acid	8.88	5.93	4.35	1.60
Glutamic acid	8.31	4.55	3.80	1.43
Arginine	7.50	—	3.27	2.21
Iminodiacetic acid	10.56	7.36	5.71	2.60
Sulfamic acid	—	—	0.85	—
Dipicolinic acid	9.10	8.70	6.36	4.36
Aminomethylphosphonic acid	8.12	—	—	1.71

Source: Ref. 9.

acid group. Notice that its stability with Cd is much less than for chelating ligands. Some other chelating groups of note include aminophosphonic, pyridine, and thiol groups. It should be noted that since the thiol sulfur atom coordinates so strongly with transition metals, its utility is limited to applications where interfering metal ions have been removed prior to its use for metals such as mercury. Chelating resins can also be operated in the presence of high TDS (total dissolved solids). This has lead to their extensive use in plating operations and groundwaters containing high hardness [10].

B. Capacity

A major concern with conventional and chelating ion exchange resins is their finite capacity. Since ion exchange groups are located on the internal surface of the resin, it is imperative that the resin has a sufficient surface area to allow for sufficient capacity for ion capture. Generally, ion exchange resins have an internal surface area of around 200–400 m^2/g. Assuming a functionalized group surface density of 10 groups per nm^2, the theoretical capacity of the resin would be approximately 3–6 meq/g. For example, a resin containing sulfonic acid is typically constructed by copolymerizing styrene and divinylbenzene. Upon complete sulfonation of the matrix, the maximum theoretical capacity can be calculated to be just over 5 meq/g. It should be noted, however, that most commercial resins have a reported capacity of 1–3 meq/g due to accessibility and functionalization efficiency. In general, the only way to increase this capacity is to lower the pore size of the resin. Unfortunately, this has the side effect of increasing diffusion resistance for ion transport to the resin surface. Hence, accessibility is of paramount concern, though the dependence on surface area is still evident.

This is made clearer through a calculation of the theoretical capacity. Figure 2a shows a theoretical construct of a porous sorbent material. A representative volume of 1 cm³ has been used to simplify calculation. Notice that an equilateral triangle has been used as the face of the volume to allow for the tightest packing of cylindrical pores. The overall porosity of the sorbent is determined from the design in Fig. 2b. In this calculation, R was chosen to be 20% greater than the pore radius to provide material stability. The resulting porosity is approximately 63% and is independent of pore radius. Subsequently, the material internal surface area can be calculated for various pore sizes. A representative sample of material surface areas is shown in Table 2. The theoretical capacity has been calculated in each case, assuming 10 exchange groups per nm^2.

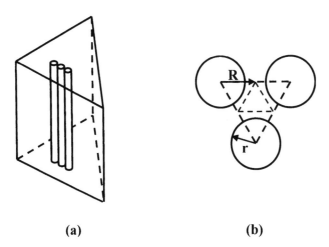

(a)   (b)

Figure 2 Schematics for calculation of microporous material internal surface area: (a) representative volume and (b) unit cell for determination of material porosity.

Table 2 Calculated Theoretical Internal Surface Area and Complementary Ion Sorption Capacities of Microporous Materials

Pore radius (nm)	Number of pores per cm^3	Surface area per cm^3 (m^2)	Surface area, $\rho = 0.7$ g/cm^3 (m^2/g)	Theoretical capacity, 10 groups per nm^2 (meq/g)
100	2.01×10^9	12.6	18	0.3
10	2.01×10^{11}	126	180	3
5	8.02×10^{11}	252	360	6
2	5.00×10^{11}	630	900	15
1	2.01×10^{12}	1262	1800	30

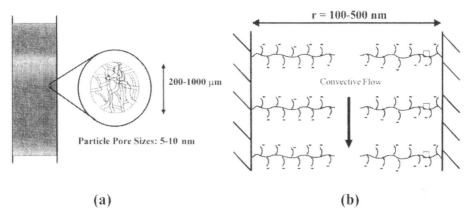

Figure 3 Schematics for (a) ion exchange resin in column operation versus (b) ion exchange by immobilized polymeric ligands in a microfiltration membrane support.

The calculation described indicates that although high capacity can be obtained by decreasing pore size, this is impractical due to site accessibility problems, kinetics, and regeneration problems. Accessibility problems develop since hydrolyzed species, such as $Pb_x(OH)_y^{(2x-y)+}$, will not be able to diffuse into very microporous sorbents. Hence, although the theoretical capacity of the resin is increased, in practice sorption is either not enhanced or is hindered. The kinetics of ion sorption also suffer since the diffusion paths are lengthened and hence column residence time is increased. Regeneration is made more difficult, as sorbed ions must diffuse out of the resin. On the other hand, the use of larger pore materials (such as microfiltration membranes) containing immobilized polyfunctional groups (Fig. 3) will overcome the above-mentioned problems as all groups are in intimate contact with solutions that flow convectively through the sorbent.

III. POLYMERIC AND MULTIPLE BINDING METAL SORBENTS

Although the mechanisms of ion exchange and chelation are well understood and have been utilized widely in industry, the matrix in which they are employed may be improved. This is mostly for the purposes of increasing capacity, improving rates of sorption, and simplifying regeneration. Bound polymeric sorbents on appropriate porous supports operate with the same basic mechanisms as ion exchange and chelation. First, since polymers in solution can be flexible (depending on the chain structure), the formation of multidentate complexes is possible (Fig. 4). An example of these flexible polyelectrolytes is polyamino acids. The amide linkage of each repeat unit on the polymer backbone provides a lone pair of electrons, while side groups, such as carboxylic acid, provide electrostatic interactions. Other examples include polyacrylic acid, sulfonated polystyrene, and quaternized polyethyleneimine.

The regular positioning of ionic groups on a polymer chain creates an additional mechanism of sorption that is not available for functional groups on a surface. Because the charged groups are in close residence (e.g., when each polymer repeat unit contains a charged group), their respective electrostatic interactions may be superimposed. Superposition of neighboring electrostatic fields is represented in Fig. 5. Notice that the combined effect is much greater than the effect of one group (inset). This phenomenon is called counterion condensation (CC). This occurs when charged groups on a polymer are close enough (<0.7 nm separation) such that their superimposed electrostatic charge fields attract and retain ions without specific interactions with any particular functional group. This critical separation distance is the so-called Bjerrum length,

$$B = \frac{e_o^2}{4kT\pi\varepsilon_0\varepsilon} \tag{1}$$

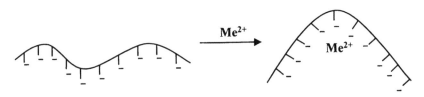

Figure 4   Polymer chain conformation to form multiple coordination sites with metals.

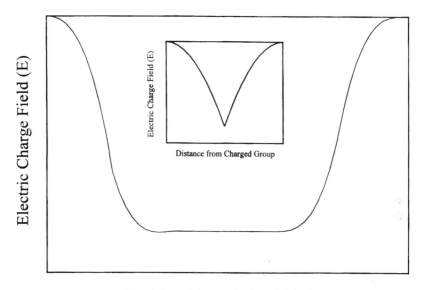

Position Along Axis of Chain

Figure 5 Superposition of electric charge fields from charged side groups of a polyelectrolyte (inset for isolated group). (Adapted from Ref. 11.)

where e_0 is the elementary charge, k is the Boltzmann constant, T is temperature, ε_0 is the vacuum permittivity, and ε is the dielectric constant of water. It should be noted that the value of B is dependent on the charge of the fixed groups. Hence, if the fixed group is divalent, then $B = 1.4$ nm.

Counterion condensation is driven by an electrostatic charge field about the polymer chains. The magnitude of this field for a single chain can be calculated using Coulomb's Law for a line charge,

$$\psi(z) = \frac{\lambda L}{2\pi\varepsilon_0 z \sqrt{4z^2 + L^2}} \tag{2}$$

where λ is the linear charge density (charge/length), L is the contour length of the polymer, and z is the radial distance from the chain. The linear charge density is controlled by ionization of the chain and hence can be altered by protonation of the chain and by specific interactions with ions in solution. The concentration of ions near the chain is controlled by the electrostatic charge field strength, as shown in Eq. (3).

$$C(z) = A \exp\left(\frac{-\varepsilon_0 \psi(z)}{kT}\right) \tag{3}$$

Notice that as the strength of the negative charge field increases, the concentration of ions near the chains increases. The theory of CC has been explained in more detail by various researchers [12–16]. It should be noted that this a nonspecific method of metal sorption, and hence the overall selectivity of the sorbent will decrease. The metal sorption capacity, however, can be much greater than for the summation of individual groups, leading to metal sorption ratios (moles metal sorbed/moles functional group) much greater than one [11].

Another important aspect of some of these polyamino acids (eg., polyglutamic acid) is the ability to form helix-coil structures. Figure 6 is a schematic showing the hydrogen bonding that promotes helix formation. When the helix is formed, the chain expands radially and is compressed axially. The result is a compressed, annular, cagelike structure that can trap sorbed species. This is critical when considering regeneration, which is generally done by increasing the ionic strength or the acidity. Figure 7 shows how a drop in pH affects helix formation in these polymers. Ionic strength and interactions with divalent cations have also been found to promote helix formation.

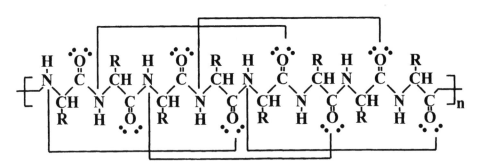

Figure 6 Hydrogen bonding between peptide amide and carbonyl groups during helix formation.

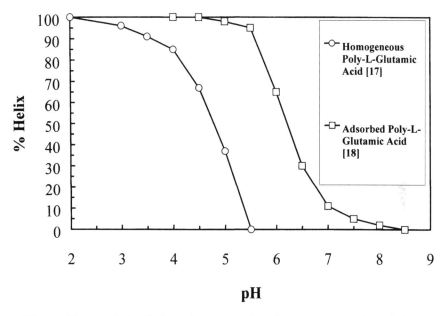

Figure 7 pH shift of helix-coil transition from homogeneous to bound phase poly-L-glutamic acid. (From Ref. 19.)

A. Polyelectrolytes for Application in Homogeneous Solution

Selectivity, though reduced somewhat by counterion condensation, is still inherent in polymeric sorbents due to the functional groups contained on the polymers. These groups are generally the same as used in ion exchange and include sulfonic acid, carboxylic acid, and amines. Much as selectivity was enhanced for ion exchange functional groups, the same can be done for polymeric sorbents. For example, polyethyleneimine can be utilized as a weakly basic anion exchanger. However, when the amine group is quaternized with bulky groups, selectivity by steric hindrances can be introduced.

Although with polyelectrolytes containing appropriate functional groups one could obtained a desired selectivity, the need for material regeneration is critical. This is because of their relative high cost, which makes recycle of the polymer imperative. The recovery of functional polymers has

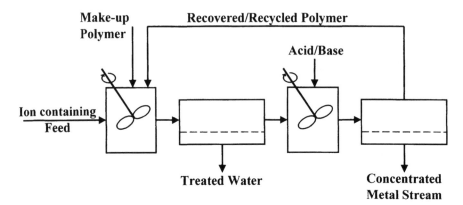

Figure 8 Polymer-enhanced ultrafiltration and recovery scheme for polymer.

been shown effectively by Scamehorn and others, as high molecular weight polymers can be separated from bulk solutions by ultrafiltration [20–23]. Once treated with high acid or base, the polymer may again be separated and recycled to the process (Fig. 8). The membrane permeate can then be further purified for ion recovery as a concentrated solution or as a salt. The addition of these separation steps, however, can add a great deal of complexity, and hence cost, to the separation process. Thus, a solid microporous support is generally more desirable, which has led to extensive research in immobilized polymeric and multiple binding sorbents.

B. Immobilized Polymeric and Multiple Binding Sorbent Materials

Functionalized materials have been long established in chromatographic processes for purification of proteins, enzymes, etc. Hence, the use of similar materials for metal sorption was a natural extension. It should be noted that this involves affinity separations (typical for protein separations), as simple surface groups are often inadequate for the desired selectivity. Therefore, the functional groups are often very complex moieties, or oligomeric structures, such that multidentate complexes may be formed. Also, as differences between metal ions are not as distinct as those between enzymes, selectivity is generally built in for various groups of metals. Since the removal, recovery, and reuse of transition metals is of immense interest, it is natural that many of the advancements in the field were made for their removal. Since the technology of functionalized cellulosic, chitosans [24], polystyrene [25], and silica materials is well understood, these materials are ideal candidates for immobilization of metal sorbing functional groups.

Cellulose is a very versatile material, and the attachment of functional groups to its surface is well understood. Functional groups may be attached to its surface through a variety of surface chemistries (e.g., Schiff-base chemistry with aldehyde groups). Table 3 shows various functional groups that have been attached to cellulose in this manner. The functional groups can vary from simple moieties, such as phosphonic acid, to more complex structures. In general, complexity is added to impart additional selectivity to the sorbent. Researchers have also attached oligomeric type groups [27], which offer more chelating properties and better selectivity for transition metals. It should be noted that the metal sorption capacities are not particularly high (0.5–1.5 meq/g), as the functionalized groups contain a relatively low number of repeat units. As will be shown with polymeric ligand functionalized membranes, when the number of repeat units is increased, the sorption capacity can be increased by an order of magnitude.

Silica gels and particles have also been widely employed as support structures for functional groups. An excellent review of silica functionalization procedures has been published in the literature [29]. In the case of silica, functionalization typically proceeds from silanol groups. Various alkoxy- and chloro-silanes are available that react with these silanol groups to form stable siloxane bonds. These are functional silanes and may con-

Table 3 Metal Sorption Capacities of Various Cellulose-Based Chelating Adsorbents

Group	Metal	Capacity (meq/g)	pH	Ref.
Phosphonic acid	Pb^{2+}	0.66	2–12	[26]
Carboxymethyl	Pb^{2+}	0.52	2–10	[26]
Triethylenetetramine pentaacetic acid	Cu^{2+}	1.5	3	[27]
8-oxa-2,4,12,14-tetraoxopentadecane	UO_2^{2+}	0.62	4.5	[28]
Carbonic acid	Cu^{2+}	0.66–1.3	4	[28]
Amidoxime and hydroxamic acid	UO_2^{2+}	0.56	4.5	[28]

tain a variety of end groups, including glycidyl, amine, thiol, and carboxylic acid groups, to name a few. These materials may then be reacted with various chelating agents to impart metal sorption capacity. Tien and Chau [30] have shown an excellent capacity for Cu(II) of 3 meq/g by incorporating amine groups during sol-gel formation of the sorbent. Lee et al. [31] achieved a lower sorption capacity for cyanide complexes with Cu(I) by attaching oligomeric structures. Deorkar and Tavlarides [32] have also successfully achieved Sb(III) sorption (0.53 meq/g) by immobilization of pyrogallol moieties on silica. Note that in each case, the same maximum capacities for ion exchange materials have still not been surpassed, due mostly to the persistent dependence on surface area.

   For both cellulose and silica supports, multiple binding capability has been imparted to an essentially inert material. Operation with these sorbents is typically carried out in column operation. Therefore, although many of the diffusion resistances that were encountered with microporous supports have been minimized, large pressure drops will still be an issue. Also, there are flow rate limitations to prevent channeling through the bed. Some of these issues can be overcome by changing the mode in which the feed solution contacts the material. In general, this has been accomplished by forming the material into a membrane conformation where a large area containing many functional groups may be packaged into a relative small unit, and additional flow can be treated by increasing the number of modules.

C. Membrane-Based Substrates for Polymeric and Other Metal Binding Functional Groups

Membrane support characteristics needed for immobilized (covalently) monomeric to oligomeric to polymeric ion exchange/chelate groups consist of ease of functionalization, accessibility, ease of regeneration, possible acid/base stability, potential for convective flow operation (i.e., MF membranes), adequate surface area, and scalability. For metal separation, various types of membrane geometries have been used and reported in the literature (Fig. 9). It is well known that separation of metals by dense, nonfunctionalized membranes (Fig. 9a) relies on the difference of diffusive transport rates such as those found in reverse osmosis (RO), and thus

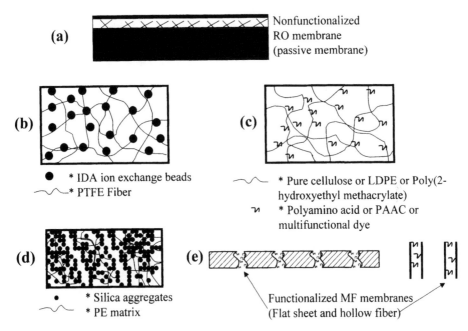

Figure 9 Membranes for ion separation: (a) conventional RO membrane, (b) membranes containing ion exchange beads, (c) fibers and other polymeric matrices functionalized with various ligands, (d) PE-Silica composite MF membrane, and (e) functionalized flat sheet and hollow fiber MF membranes.

membranes need to be operated at very high pressure (30–80 bar) to get appropriate flux. Although these membranes are tight, they do remove monovalent salts, such as NaCl, in addition to their highly effective rejection (>99.9%) of divalent metals. On the other hand, when the primary purpose is creation of a sorbent/ion exchange device, particularly for transition metals and other toxic anions, the main objective is to maximize capacity (metal sorption) at high throughput rate.

Membrane-based sorption devices can take the forms of fibrous mats containing ion exchange beads (Fig. 9b), directly functionalized polymers containing multiple binding functional groups (Fig. 9c), direct incorporation of particles (e.g., silica) in a polymeric microfiltration matrix (Fig. 9d), which can be postfunctionalized with metal binding sites, and traditional microfiltration membranes that can easily be postfunctionalized

(Fig. 9e). The ability to operate a membrane-based sorbent device under convective flow mode will have considerable advantages in terms of kinetics, accessibility, and throughput rate. The schematics shown in Figure 9c–e fall in this category and will be discussed further. Although various materials can be used for functionalization supports, cellulose and silica-based materials can be functionalized by well-established methods in the literature. Polyolefin-based materials have also been studied extensively, but these require irradiation prior to functionalization.

In an effort to improve on the performance of these membranes, particularly for metal sorption applications, more hydrophilic polymers have been used to increase swelling and hence flux properties [33]. In this work, alginate and cellulose were combined in a membrane matrix for strontium and cadmium sorption (~0.75 meq/g). Cellulose was used to provide structural support, while the alginate provided carboxylic acid and hydroxyl groups for metal sorption. Swelling of the membrane was significant, such that a membrane with a dry pore radius of 7 nm could have a flux of 2.4×10^{-4} cm^3/cm^2/s with a transmembrane pressure of 1 bar. It should be noted that the membrane swelled immensely, with greater than 60% water content in the swollen membrane.

These membranes do not provide the selectivity required for mixed ion streams. Therefore, other researchers have tried to achieve the desired selectivity by incorporating ion exchange resin in an inert polymer matrix. A representation of these sorbents is shown in Fig. 9b. Notice that the inert material only provides a matrix to hold the resin in place. The result is a thin film through which a feed solution containing a contaminant may be permeated. This is especially useful when the feed solution contains a large amount of suspended solids that would clog a standard fixed bed. An excellent overview of this technology—composite ion exchange material (CIM)—has been published by Sengupta and SenGupta [34]. The typical properties of a CIM are 90% chelating resin containing IDA groups and 10% PTFE, membrane thickness 0.4–0.6 mm and a nominal pore size of 0.4 μm. It should be noted that, depending on the polymer selected, the inert material may also be functionalized [35]. In their work, a mixed nonwoven fabric of polyethylene and polypropylene was radiation-induced grafted with glycidyl methacrylate groups. Subsequent sulfonation of the material created a fairly open structure with sulfonic acid functionality.

Polymeric functionality may also be incorporated in a membrane morphology by polymer blending. In general, an inert material is used to provide structural stability to a less robust, though more functionalized, polymer. Park et al. [36] have presented a chelating membrane composed of poly(vinyl alcohol) and poly(N-salicylidene allyl amine) for Co(II) sorption. In this work, the goal was to set up stable complexes in the membrane to protect the easily oxidized groups of the polyamine. Cobalt was complexed in the membrane to an extent as high as 15% by weight. Unfortunately, the membrane was found to become too brittle as the cobalt content increased beyond that threshold. Other work with polymer blending has been done by incorporation of polyaniline in a cellulose acetate membrane [37].

It should be noted that a change in morphology to membrane-based systems does not remedy the surface area dependence of most ion exchange and chelation materials. Some diffusional resistances have been minimized, though without sufficient capacity the utility of these materials is somewhat limited. Therefore, researchers have looked to the immobilization of polymeric structures in the free pore space of membranes so that intimate contact can be made with the sorbent material while the functional groups are tethered away from the surface. Hence, these structures need only to be anchored on the surface, and multiple functional groups can originate from the same surface site.

D. Graft-Polymerized Membranes

Graft polymerization has also been successfully achieved with hollow fibers. The hollow fiber construct is desirable, as it allows for the greatest surface area to volume ratio in a packaged module. It should be noted, however, that irradiation of these membranes results in a higher concentration of radicals near the pore mouths, and not all of the internal surface area is utilized fully (Fig. 10). In the case of polyethylene hollow fibers, radiation grafting has been used to functionalize the material with glycidyl methacrylate. Chelation or ion exchange groups may then be added to the material to provide functionality for ion sorption. A great deal of work has been done in this area by researchers in Japan including the incorporation of IDA functionality [38,39] and diethylamine groups [40]. For example, Konishi et al. [38] and Li et al. [39]

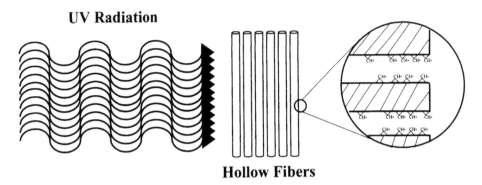

Figure 10 Radiation-induced free radical formation on hollow fibers.

have found sorption capacities of 3.4 meq/g Co(II) and 0.94 meq/g Pd, respectively.

Polymeric groups may also be functionalized on these graft-polymerized membranes. For example, researchers have followed graft polymerization of glycidyl methacrylate with ethylenediamine to impart chelation groups to the membrane [41]. Since the substrate is a microfiltration membrane, the pore spaces (0.1–1 μm range) are quite large compared to ion exchange resins, and a large amount of polymeric groups may be incorporated into the membrane (recall Fig. 3). Monomers may also be employed that already contain functional groups. One example of this is graft polymerization of polyvinylpyrrolidone on silica membranes [42]. In this case, since silica is rather easy to functionalize by silanization chemistry, irradiation of the support was not necessary. Following silanization of the surface, vinyl pyrrolidone was polymerized in situ to form a brush structure in the membrane pores. Researchers have also immobilized poly-(acrylic acid) on polyvinylidene fluoride membranes [43].

Membranes may also be graft-polymerized with a combination of monomers. In this fashion, different chelating functionalities can be designed into a sorbent. Work in this area has been published [44,45] for the purpose of treating wastewater. In one case, acrylic acid was graft-polymerized with styrene. Post-treatment of the grafted membrane (LDPE) by sulfonation and alkali allows for incorporation of both sulfonic acid and carboxylate groups. The same type of membrane has also been graft-polymerized with acrylic acid and vinyl pyridine to incorporate chelating groups. These membrane sorbents have shown excellent metal sorption capacities of 2 meq Cd/g and 13 meq Pb/g for poly(acrylic acid) functionalized polyethylene membranes. It should be noted that all metal sorption experiments were performed by soaking the membrane. Metal uptake was a function of degree of grafting, which in turn affected the water uptake of the membrane. The presence of sulfonated styrene groups in general lowered the overall capacity as carboxylic acid groups are more efficient on a capacity versus mass basis. However, when the membrane was co-polymerized with 2-vinyl pyridine, rather than styrene, the capacity was found to increase for all examined metals due to enhanced chelation of metal by the pyridyl moiety. This group reversed the metal sorption capacities for Cd and Pb, resulting in sorption values of 6 meq/g and 5 meq/g, respectively.

E.   Incorporation of Multiple Binding Dye Ligands

Multiple binding sorbent structures (molecules containing varied, multiple exchange groups) may also be immobilized in membrane structures. A great deal of work has been done on the impregnation of membrane structures with multifunctional dyes, such as Alkali Blue 6B and Cibacron Blue F3GA. These types of dyes have also been used for affinity bioseparations [46]. Denizli et al. [47,48] have done extensive work with immobilized dyes. The advantage of dye impregnation is that a large number of different functional groups can be incorporated in the same membrane sorbent. Figure 11 shows schematics of Alkali Blue 6B and Cibacron Blue. Notice how interactions with ions can take place with many different groups, including amines, sulfonic acids, amino linkages, and aromatic nitrogens.

The membranes used were all macroporous, hydrogel films with high water content. The two films used were poly(2-hydroxyethyl methacrylate) (HEMA) and polyvinylbutyral. Equilibrium swelling ratios ranged between 58 and 89%. The multiple binding dyes were covalently attached to the films. The metal pick-up rate was quite rapid (80% of equilibrium capacity in less than 15 minutes) with these hydrogel films, and the equilibrium capacity ranged from 8 mmol/m^2 film to 65 mmol/m^2 film. For the case of HEMA films containing Alkali Blue, the equilibrium capacity corresponded to 2.8 mmol Cd/mmol dye and 4.3 mmol Pb/mmol dye.

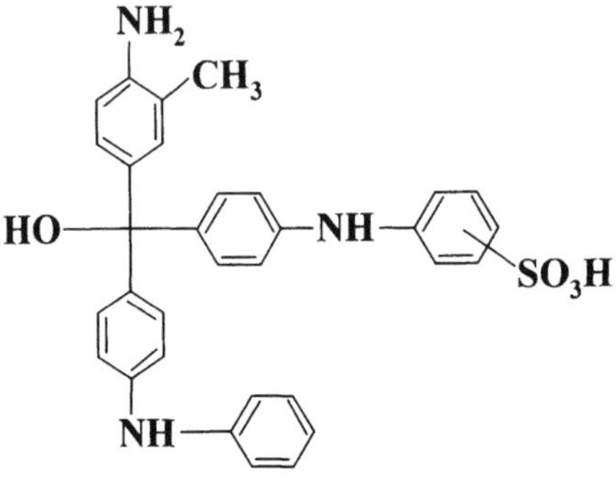

(a)

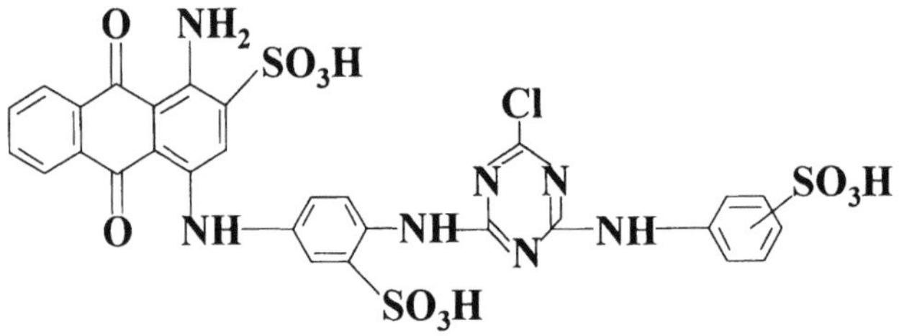

(b)

Figure 11 Dye chemical structures of (a) Alkali Blue 6B and (b) Cibacron Blue F3GA used in metal binding.

For both dyes the selectivity sequence is $Pb^{2+} > Zn^{2+} > Cd^{2+}$. The regeneration was highly effective with 0.1 M HNO_3 and for all three metals greater than 90% regeneration was reported.

IV. POLYAMINO ACID FUNCTIONALIZED MICROFILTRATION MEMBRANE SORBENTS

It has already been shown that radiation-induced graft-polymerization is an effective technique for immobilization of polymeric structures in a membrane. However, as observed with ion exchange and chelation resins, to engineer in selectivity, simple polymeric structures such as polyacrylic acid are insufficient. Even copolymerization of multiple groups [44,45] can only offer limited flexibility in designing truly selective membrane sorbents. Therefore, it is critical to examine methods of immobilizing fully polymerized chelates and polypeptides into membranes. One particularly interesting group of polymers for this purpose is polyamino acids. These offer a wide range of side groups, such as carboxylic acids, amines, and thiols, with the presence of a single terminal amine. The presence of this terminal group provides excellent opportunities for single point attachment of these polymers through the appropriate chemistry, and thus avoiding crosslinking and loss of functionalized sites.

However, before examining the attachment of these polymers, first the membrane substrate must be considered. This is because the membranes must have sufficient surface area for immobilization, high porosity, and microfiltration-type pores. Microfiltration (MF) membranes (pore sizes: 0.1–1 μm) are ideal, as they can easily accommodate large (1000–50,000 MW) polymers with adequate surface area for attachment ($\sim$10–150 m^2/g). The thickness of membranes used for our investigations ranged between 100–200 μm. This corresponds to the availability of 8–400 exchange sites (repeat units) per membrane surface site. Recall from Table 2 that a material with 0.1 μm pore size would have a theoretical capacity of only 0.3 meq/g involving monomeric ligands. However, the number of metal binding sites can be increased by as much as 100 times after functionalization with a 100–repeat unit polymer. Materials of construction can vary widely, though our work has been done with cellulose acetate [19], pure cellulose mats, and silica-based [11] membranes due to ease of functionalization.

A. Membrane Functionalization and Polyamino Acid Immobilization

Figure 12 shows the structures and functional groups of polyamino acids used in our studies at the University of Kentucky. The terminal amine group can easily be utilized for functionalization. In the case of cellulosics, the goal is to generate aldehyde groups on the surface. This allows for simple Schiff-base chemistry for reaction with a terminal amine. When the starting material is cellulose acetate, the acetate groups must first be

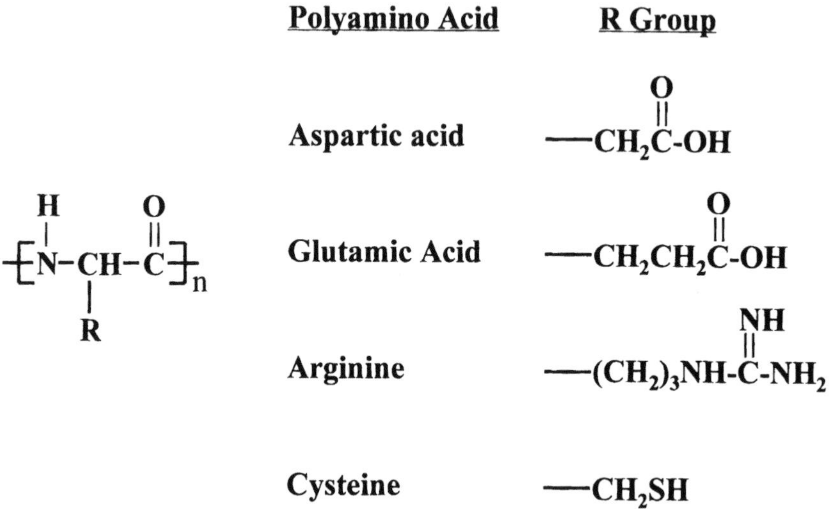

Polyamino Acid   R Group

Aspartic acid

$$—CH_2\overset{\overset{\displaystyle O}{\|}}{C}\text{-OH}$$

Glutamic Acid

$$—CH_2CH_2\overset{\overset{\displaystyle O}{\|}}{C}\text{-OH}$$

Arginine

$$—(CH_2)_3NH\text{-}\overset{\overset{\displaystyle NH}{\|}}{C}\text{-}NH_2$$

Cysteine

$$—CH_2SH$$

$$\begin{array}{c} H \quad\quad O \\ | \quad\quad\; \| \\ {\footnotesize\left[\!\!\right.}N\text{-}CH\text{-}C{\footnotesize\left.\right]_n} \\ | \\ R \end{array}$$

Figure 12 Generalized structure of polyamino acids with representative side groups.

hydrolyzed. Subsequently, when hydroxyl groups are present in adjacent positions of the carbohydrate repeat units, the rings may be opened by oxidation to get aldehyde groups. Oxidation can be achieved by sodium periodate, for example. This process is shown schematically in Fig. 13. On the other hand, with pure cellulose as a starting material, the hydrolysis step is not required. With pure cellulose the resulting material is in a mat form (thin fibers 30–50 nm diameter with high aspect ratio) rather than as a conventional microfiltration membrane.

Immobilization on silica supports proceeds similarly to that reported for pyrrolidone functionalized membranes. In our work, on the other hand, the base unfunctionalized microfiltration membrane was a commercially available polyethylene–silica composite (PE-Silica). Silica composite membranes can easily withstand high concentration acids (10–30% sulfuric acid) and have good resistance to organic solvents. The silica surface is reacted with a silane containing an amine or glycidyl group. In the case of amine functionality, glutaraldehyde can be used to crosslink the polymer to the surface. Otherwise, the amine groups can react directly with the glycidyl moiety to form a stable single bond. A schematic of this procedure

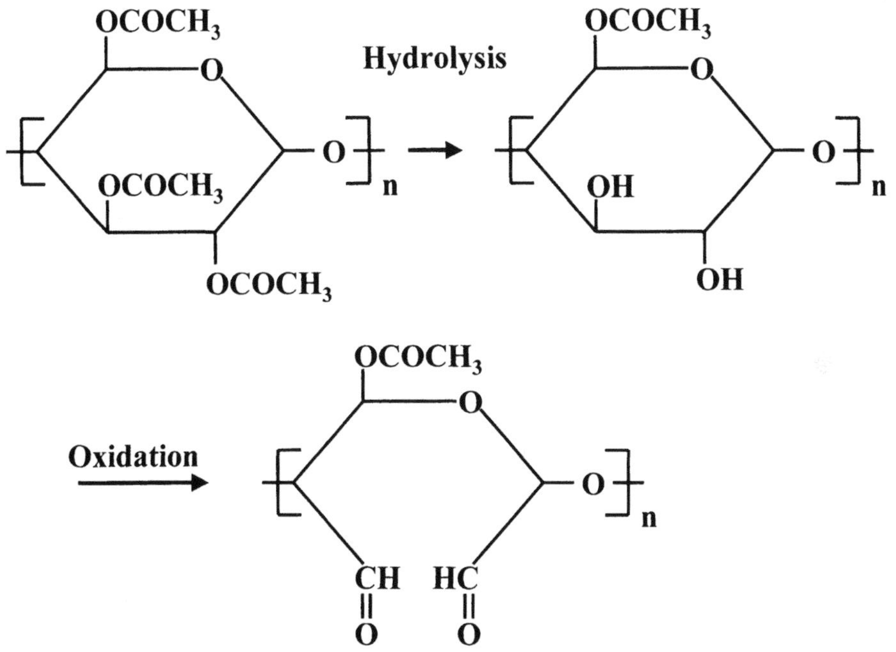

Figure 13   Derivatization scheme for cellulose acetate membranes.

is shown in Fig. 14. Notice that reaction with the polyamino acid terminal amine yields an acid stable single bond. This is in contrast to cellulosic, aldehyde functionalized materials discussed previously, where Schiff-base chemistry results in a double bond, and hence a reducing agent must be employed to impart acid stability to the sorbent. It should be noted that aldehyde/epoxide derivatization and polyamino acid functionalization of the sorbents have been performed using convective flow processes.

When the membrane support has been functionalized, it will resemble the schematic shown in Fig. 3b. Notice that the polymeric ligand structures are attached at one end only. Hence, the bulk of the polymer is free to dangle in the pore and interact with ions that flow convectively through the pore. Typical water fluxes for these functionalized membranes are 10 to 50 $\times$ 10^{-4} cm^3/cm^{2}s at 0.2–5 bar. This corresponds to sorbent residence times as low as eight seconds, which are much too low for diffusional processes. Hence, convective transport of the ions makes diffusional

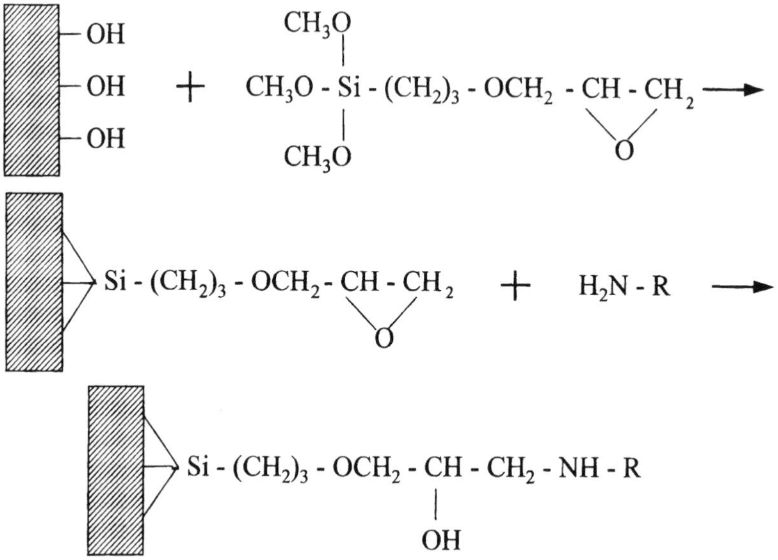

Figure 14 Preparation of silica composite membranes by silane derivatization followed by polyamino acid functionalization.

paths of transport negligible. Ion sorption and desorption during regeneration are hence much faster than observed for conventional ion exchange processes.

B. Metal Sorption and Ion Exchange Mechanisms and Capacities

All experimental results reported here were done with metal nitrate salts. Equilibrium capacity of various metals by polyamino acid functionalized membranes is a function of the type and concentration of metal, the number of repeat units of the functionalized polymer, and the degree of functionalization. Polyamino acid (i.e., PLGA, polyaspartic acid, etc.) structure, helix-coil transition, and degree of ionization play an important role in metal interactions. However, to understand the extent of interactions with specific metal ions (e.g., Cu^{2+} versus Ca^{2+}), an understanding of normal chelation mechanisms with monomeric chelates (e.g., aspartic acid) is useful. An excellent review of metal chelation and metal–ligand correlations has been reported by Hancock and Martell [49]. These correlations include interactions of metals with negatively charged oxygen donors, saturated and unsaturated nitrogen donors, etc. Figure 15 represents the complexation constants of various metals versus metal ion acidity (first metal hydrolysis constant) for aspartic acid and amine-free malonate. Metal–ligand stability constants increase with the acidity of the metal for both cases. The metal selectivity on carboxylic acid–based exchangers is expected to follow $Cu^{2+} >> Ca^{2+}$. This was indeed found to be true in our experiments.

On the other hand, polyamino acids, in contrast to monomeric acids, present another method of sorption by counterion condensation. There are three primary mechanisms for metal sorption. These are ion exchange, chelation, and electrostatic (in condensation zone) binding. The

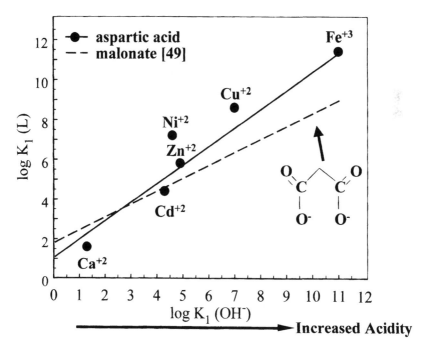

Figure 15   Effect of metal ion acidity on metal–chelate stability.

first two mechanisms occur in conventional ion exchange and chelation resins. The last mechanism is a function of the polymeric nature of the ligands, and sorption by this technique is referred to as counterion condensation. This final mechanism is partly responsible for the large binding capacities observed with membrane-based sorbents. The role of charge spacing on the polymeric chain has been discussed previously. However, an additional effect occurs when the charged polymer chains are in a bound phase, particularly in a pore. Figure 16 indeed demonstrates the role of counterion condensation as evidenced by the molar metal/COOH ratio much greater than 1 for the pore-bound polyamino acid. For the case of silica particles, the surface to which the polyamino acids are attached is nonporous, and hence the chains are not in close residence, as they would be in the microfiltration membrane pores. High sorption ratios were also observed with other types of microfiltration membranes, and the selectivity sequence (Fig. 17) followed the order $Cu^{2+} \gg Pb^{2+} > Cd^{2+}$. The enhanced copper capacity may also be due to high affinity for the polyamino acid amide linkages.

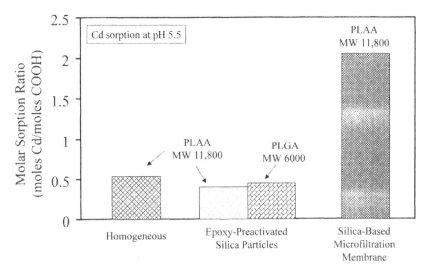

Figure 16 Surface conformation effect on polyamino acid functionalized silica-based sorbent metal capacity (PLAA: poly-L-aspartic acid, PLGA: poly-L-glutamic acid). (From Ref. 11.)

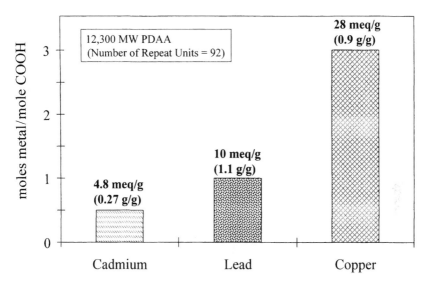

Figure 17   Pb, Cu, and Cd sorption on poly-(α,β)-DL-aspartic acid functionalized cellulose acetate composite (Osmonics) membranes. (From Ref. 11.)

Up until now, we have discussed the use of carboxylic-based interactions. However, the use of amine-based ligands, such as polyarginine, should allow sorption/ion exchange of oxyanions. This phenomenon was tested with experiments conducted using pure cellulose membranes. Table 4 shows the metal sorption results obtained with polyarginine. It can be seen that we indeed were able to attach polyarginine for anion pick-up. Although the polymer loading was not optimized, it is interesting to observe that the molar sorption ratio (moles metal/moles repeat unit) is approximately 1. In contrast, PLGA (carboxylic groups), which is not ionized at pH 3.5 and is in complete helical form, resulted in negligible pick-up of Cd^{2+}.

In reference to membrane materials, it should be noted that the derivatization steps required to form aldehyde or epoxide groups are different for cellulosics and silica. In addition, the PE-Silica and pure cellulose membranes have internal surface areas approximately ten times greater than cellulose acetate materials. As a result, aldehyde/epoxide groups ranged between 1 to 7×10^{-4} mmol/cm^2 external membrane area. Figure 18 shows the metal sorption capacity ($\sim$500 mg/L solution, pH 5–5.5) on the different types of membranes functionalized with polyglutamic

Table 4 Cellulosic-Membrane Bound Polyarginine (12,000 MW, ~85 repeat units) and PLGA (14,000 MW, ~90 repeat units) for Metal Sorption at Low pH

Metal type	pH	Membrane material	Polyamino acid	Grams metal per grams membrane	Aldehyde (mmol/cm^2)	Polyamino acid (mmol/cm^2)	Moles metal per moles repeat unit
Cr (as Cr$_2$O$_7^{2-}$)	4.3	Pure cellulose	Polyarginine	0.15	6.3×10^{-4}	0.64×10^{-4}	0.9
Cd*	3.5	Modified cellulose acetate	PLGA	<0.01	1.3×10^{-4}	0.54×10^{-4}	—

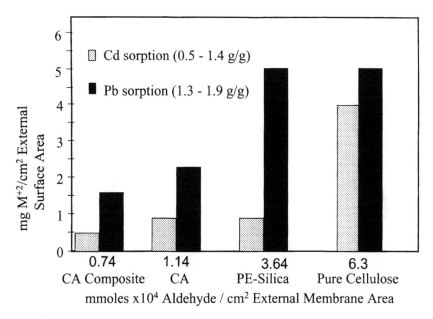

Figure 18 Comparison of lead and cadmium sorption capacities with various membranes (polyamino acid functionalized).

acid and polyaspartic acids of different molecular weights [50]. In ion exchange terminology, the capacities demonstrated by these membranes are as high as 30 meq/g (compared to 1–3 meq/g for a typical ion exchanger). All three of the membrane materials shown in Fig. 18 have their advantages. The pure cellulose has the highest capacity by weight (prepared in mat form), PE-Silica has the best acid stability and tolerance to harsh environments, and the cellulose acetate membranes are very widely available. The extent of sorption is a function of the amount of polyamino acid present in the membrane, accessibility, and the concentration of the metal solution.

C. Typical Breakthrough Curves and Selected Applications

The usefulness of polyamino acid functionalized membranes, in terms of capacity, role of material type, etc., has been discussed herein. Extensive experiments conducted both in recycling and continuous mode showed that a high rate of sorption can be obtained under convective flow conditions. Figure 19 shows a breakthrough curve for a 1000 MW polyaspartic acid (Na form) functionalized PE-Silica membrane. In order to avoid variations that can be introduced by using a commercial polyamino acid, a specific octapeptide was synthesized at the University of Kentucky. As expected, Na was released by the membrane as Cd was sorbed. A material balance showed a ratio of 2 moles Cd sorbed per mole Na released. This demonstrates the presence of counterion condensation and how this is an important mechanism of metal sorption involving bound polyelectrolytes. To further demonstrate metal selectivity under continuous operation mode, experiments were conducted with 6000 MW PLGA (~40 repeat units) using a feed mixture of Cd^{2+} and Pb^{2+} (Fig. 20). As expected from metal acidity, the breakthrough curve follows the sequence $Pb^{2+} > Cd^{2+}$. It is important to observe that using functionalized microfiltration tech-

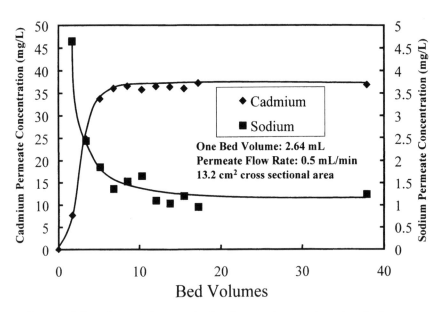

Figure 19 Charge balance and role of counterion condensation for low concentration (50 mg/L) cadmium sorption on octapeptide polyaspartic (1000 MW) acid functionalized polyethylene–silica membranes.

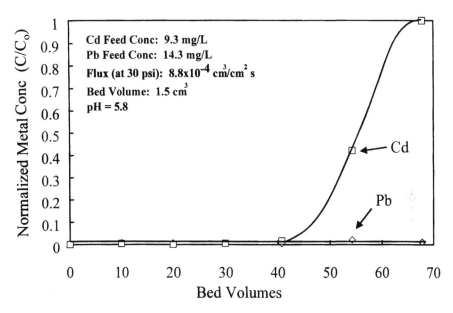

Figure 20 Competitive sorption of Pb and Cd by polyglutamic acid (6000 MW) functionalized polyethylene–silica composite MF membrane.

nology, one could separate and recover valuable metals such as Cd from Pb.

Metal recovery can be achieved in a variety of ways. First, as noted, when the sorbent is sufficiently selective, a mixture can be separated when one component is retained on the sorbent while another passes through to the permeate stream. The retained metal may then be recovered by regeneration of the sorbent. In conventional ion exchange, regeneration is simply by reversal of the equilibrium sorption reaction. On the other hand, with bound polyamino acid ligands, in addition to the same reversal phenomenon, one could take advantage of helix-coil transitions to obtain selective regeneration. In general, regeneration is done by acid and/or salt treatment. Table 5 shows some example Cd and Pb regeneration results for polyamino acid functionalized membranes. Recall from Fig. 7 that around pH 4 significant helix formation exists, for example, with polyglutamic acid. One may expect that metals with a high acidity (i.e., Pb, Fe, etc.) will be strongly retained in the helix region, unless the pH is signifi-

Table 5 Regeneration of Polyamino Acid Functionalized Membrane-Based Sorbents

Regeneration solution	Membrane	Metal sorbed	Mass of metal sorbed (mg)	Mass of metal recovered (mg)	Percent recovery of metal
20 wt% NaNO₃ @ pH 3	PE-Silica	Cd	0.66	0.56	80
20 wt% NaNO₃ @ pH 3	Cellulose-acetate composite	Cd	3 ± 0.1	3 ± 0.1	~100
20 wt% NaNO₃ @ pH 3	Cellulose-acetate composite	Pb	13.9	2.3	16
20 wt% NaNO₃ @ pH 7	PE-Silica	Pb	18–25	1–5	4–18

cantly lowered to completely protonate the COO^- groups. Table 5 indeed shows that with two different membranes, Cd recovery under mild (pH 3) condition is very high (80–100%). On the other hand, most Pb is retained at the same regeneration conditions, and thus this approach provides an intriguing possibility of selective metal recovery. Of course, as expected, at pH 7 no regeneration is possible.

The applications for functionalized MF membrane sorbents can vary widely, from metal removal in complex streams (containing hardness and competing transition metals) to supplemental treatment for RO and nanofiltration (NF) retentate and permeate streams and complete encapsulation of highly toxic (Hg, Pb) or radioactive metals. In metal finishing operations, one could get mixtures of metals and nontoxic inorganics. Figure 21 shows our results for staged separation of Cr(III) from an actual Department of Defense wastewater. Notice that competing ions, such as Cu^{2+}, Pb^{2+}, and Zn^{2+}, and a relatively high (0.2 wt%) concentration of Na are present in the stream. However, since the stream has been previously

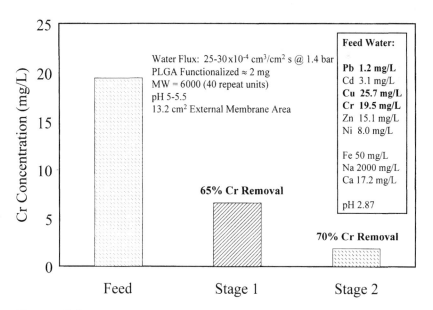

Figure 21 Cr removal from a multiple metal containing stream (actual wastewater) by staged functionalized microfiltration membrane sorbent operation.

reduced, Cr(III) is present as Cr^{3+}, and hence is selectively sorbed. Also, staged operation showed an overall Cr(III) removal of 90% in 2 single pass stages for PLGA functionalized PE-Silica membrane. It should be noted that the same result could be obtained by increasing the membrane thickness, which would allow for greater polyamino acid functionalization. Sorbent selectivity was found to be $Cr^{3+} > Pb^{2+} >> Cu^{2+} > Cd^{2+}$, Zn^{2+}, and Ni^{2+}.

Two additional example applications for this technology are shown in Figs. 22 and 23. In each case, functionalized MF membrane sorbents have been incorporated in hybrid processes with reverse osmosis and nanofiltration. In the case of RO, although rejections of dissolved metals and oxyanions is excellent (> 98%), a large retentate stream is also produced (up to 10% of feed volume) that is concentrated in the rejected species. Often, this stream is insufficiently concentrated for direct recycle (e.g., metal plating operations) and hence must be further concentrated. Functionalized MF membrane sorbents are an excellent supplemental process for this application as their large throughput rate (due to convective

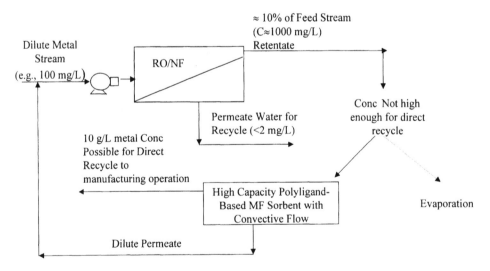

Figure 22   Example of hybrid reverse osmosis/polyligand functionalized MF membrane system for production of high quality permeate and concentrated stream.

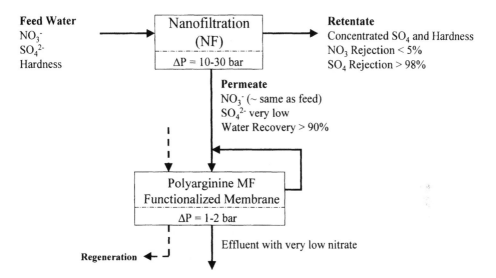

Figure 23 Example of hybrid nanofiltration/polyligand functionalized MF membrane system for nitrate reduction from water.

flow operation) allows for treatment of large volumes, while ease of regeneration makes production of a concentrated regenerant stream possible. As shown in Fig. 22, the generation of a 10 g/L regeneration stream allows for use of technology such as electrowinning for metal recovery.

Another application for these sorbent materials is removal of nitrate from drinking water. It is well known that high nitrate concentrations (> 10 mg/L) are toxic. Recall from Fig. 1 that selective ion exchange resins are available for nitrate removal from sulfate-containing streams. Unfortunately, the use of these resins may be cost prohibitive, particularly for high volumes and for streams containing high TDS, such as hardness. Figure 23 shows an example scheme for hybrid operation of NF with functionalized MF membrane sorbents. Charged NF membranes have SO_4^{2-} rejections $>>> NO_3^-$. Hence, NF can be used effectively for separation of sulfate (which will compete for sorption sites) and hardness. High throughput treatment of the permeate stream can then be accomplished with amine (polyarginine) functionalized MF membrane sorbents.

Polyligand functionalized MF membranes, because of their very high capacity (> 1 g/g), are particularly suitable for applications involving encapsulation of toxic wastes without the use of regeneration. This will significantly decrease the volume of the hazardous waste (conversion to a low volume solid waste that can be easily disposed). The metal wastes that could be treated this way are streams that have no significant resale value, but are still considered hazardous, such as lead, dilute radioactive wastes, etc. For this type of application, the functionalized sorbent would have a significant advantage over RO or ion exchange because of the high volume reduction factor and the conversion of the stream to a solid waste which can be disposed of in a secured disposal site. Using our experimental data for Pb sorption on PLGA functionalized MF membranes, it was calculated that six barrels of dilute wastewater (e.g., 40 mg/L) could be encapsulated in a 300 ft^2 membrane area which is a typical size of a spiral wound element. Thus, the functionalized MF membranes provide an effective option to off-site liquid waste disposal.

ACKNOWLEDGMENTS

The authors acknowledge NSF-IGERT, Eastman Chemical Co., and the US EPA for the financial support of this project. The authors thank Dr. Jamie Hestekin for some of the results on cellulosic materials. The authors also acknowledge the significant contributions of Dr. L. G. Bachas, Dr. S. K. Sikdar of the US EPA, and Dr. S. Kloos of Osmonics for providing valuable suggestions throughout the course of this work.

REFERENCES

1. T Kojima, T Takano, T Komiyama. J Membr Sci 102:49–54, 1995.
2. FG Helfferich. Ion Exchange. New York: McGraw-Hill, 1962.
3. AK SenGupta, ed. Ion Exchange Technology. Lancaster, PA: Technomic Publishing, 1995.
4. SD Alexandratos, DW Crick. Ind Eng Chem Res 35:635–644, 1996.
5. RA Beauvais, SD Alexandratos. React Funct Polym 36:113–123, 1998.
6. Metals Adsorption Workshop, US EPA, Cincinnati, OH, May 1998.
7. GA Guter. In: AK SenGupta, ed. Ion Exchange Technology. Lancaster, PA: Technomic Publishing, 1995, pp 61–113.
8. S Liang, MA Mann, GA Guter, PHS Kim, DL Hardan. J Am Water Works Assoc 91:79–91, 1999.

9. AE Martell, RM Smith, RJ Motekaitis. Critically selected stability constants of metal complexes database. NIST Standard Reference Database 46, Version 2.0, 1995.
10. PS Myers. How chelating resins behave. Metals Adsorption Workshop, US EPA, Cincinnati, OH, May 5–6, 1998.
11. SMC Ritchie, LG Bachas, T Olin, SK Sikdar, D Bhattacharyya. Langmuir 15:6346–6357, 1999.
12. GS Manning. J Chem Phys 51:924–933, 1969.
13. GS Manning. J Chem Phys 51:934–938, 1969.
14. S Nilsson, W Zhang. Macromolecules 23:5234–5239, 1990.
15. JA Marinsky, MM Reddy. J Phys Chem 95:10208–10214, 1991.
16. F Oosawa. Polyelectrolytes. New York: Marcel Dekker, 1971.
17. T Kurotu. Inorg Chim Acta 191:141–147, 1992.
18. E Pefferkorn, A Schmitt, R Varoqui. Biopolymers 21:1451, 1982.
19. D Bhattacharyya, JA Hestekin, P Brushaber, L Cullen, LG Bachas, SK Sikdar. J Membr Sci 141:121–135, 1998.
20. A Tabatabai, JF Scamehorn, SD Christian. J Membr Sci 100:193–207, 1995.
21. Y Uludag, HO Ozbelge, L Yilmaz. J Membr Sci 129:93–99, 1997.
22. CS Dunaway, SD Christian, EE Tucker, JF Scamehorn. Langmuir 14: 1002–1012, 1998.
23. BF Smith, TW Robison, GD Jarvinen. ACS Symp Ser 716:294–330, 1999.
24. Y Baba, K Masaaki, Y Kawano. React Funct Polym 36:167–172, 1998.
25. RMC Sutton, SJ Hill, P Jones. J Chromatogr A 739:81–86, 1996.
26. AM Naghmush, K Pyrzynska, M Trojanowicz. Talanta 42:851–860, 1995.
27. P Burba, JC Rocha, A Schulte. Fresenius' J Anal Chem 346:414–419, 1993.
28. HJ Fischer, KH Lieser. Fresenius' J Anal Chem 346:934–942, 1993.
29. JF Biernat, P Konieczka, BJ Tarbet, JS Bradshaw, RM Izatt. Sep Purif Methods 23:77–348, 1994.
30. P Tien, LK Chau. Chem Mater 11:2141–2147, 1999.
31. JS Lee, NV Deorkar, LL Tavlarides. Ind Eng Chem Res 37:2812–2820, 1998.
32. NV Deorkar, LL Tavlarides. Hydrometallurgy 46:121–135, 1997.
33. L Zhang, J Zhou, D Zhou, Y Tang. J Membr Sci 162:103–109, 1999.
34. S Sengupta, AK SenGupta. React Funct Polym 35:111–134, 1997.
35. M Kim, M Sasaki, K Saito, K Sugita, T Sugo. Biotechnol Prog 14:661–663, 1998.

36. CK Park, MJ Choi, YM Lee. Polymer 36:1507–1512, 1995.
37. S das Neves, MA De Paoli. Synth Met 96:49–54, 1998.
38. S Konishi, K Saito, S Furusaki, T Sugo. Ind Eng Chem Res 31:2722–2727, 1992.
39. GQ Li, S Konishi, K Saito, T Sugo. J Membr Sci 95:63–69, 1994.
40. K Sunaga, M Kim, K Saito, K Sugita, T Sugo. Chem Mater 11:1986–1989, 1999.
41. S Tsuneda, T Endo, K Saito, K Sugita, K Horie, T Yamashita, T Sugo. Macromolecules 31:366–370, 1998.
42. RP Castro, Y Cohen, HG Monbouquette. J Membr Sci 84:151–160, 1993.
43. R Telaranta, JA Manzanares, K Kontturi. J Electroanal Chem 464:222–229, 1999.
44. ESA Hegazy, H Kamal, N Maziad, AM Dessouki. Nucl Instrum Methods Phys Res, Sect B 151:386–392, 1999.
45. ESA Hegazy, HAA El-Rehim, HA Shawky. Radiat Phys Chem 57:85–95, 2000.
46. JE Ramirez-Vick, AA Garcia. Sep Purif Methods 25:85–129, 1997.
47. A Denizli, D Tanyolac, B Salih, E Aydinlar, A Ozdural, E Piskin. J Membr Sci 137:1–8, 1997.
48. A Denizli, K Kesenci, MY Arica, B Salih, V Hasirci, E Piskin. Talanta 46:551–558, 1998.
49. RD Hancock, AE Martell. Chem Rev 89:1875–1914, 1989.
50. D Bhattacharyya, JA Hestekin, SMC Ritchie, LG Bachas, Functionalized microfiltration membranes: A new approach to removal and recovery of dissolved heavy metals. Proceedings of the Membrane Technology Conference, BCC, Boston, December 1998.

4

Biosorption of Metal Cations and Anions

Bohumil Volesky, Jinbai Yang, and Hui Niu

McGill University, Montreal, Canada

I. INTRODUCTION

Bisorption, as considered here, is defined as a passive metal uptake process by dead microbial or seaweed biomass that concentrates metals because of its chemical makeup. The phenomenon can find an enormous scale of applications, particularly in cost-effective detoxification of metal-bearing industrial effluents. Recent studies indicate that the biosorptive metal uptake mechanism heavily relies on ion exchange. As such it is very sensitive to pH of the solution, which also affects the solution chemistry of the dissolved metallic species. Simplistic quantitative analysis of biosorption may not take all the essential factors into account. More advanced approaches are discussed in this chapter, which examines the biosorption of cations and anions from aqueous solutions. The sorption equilibrium approach discussed here provides the essential basis for the continuous flow biosorption process arrangement, which is eventually perhaps most effectively carried out in sorption columns. An even higher degree of sophistication is required to predict the performance of the dynamic sorption column arrangement.

119

The process modeling principles are demonstrated on a more complicated uranium cation biosorption case that includes occurrence of different cationic species in the solution. Simpler selected aspects of anion biosorption are demonstrated on a case of gold cyanide anionic complex biosorption, for which a different type of sorbent material is more suitable. The importance of considering the speciation of metal ions in the solution is pointed out for the case of chromium biosorption.

II. BIOSORPTION OF CATIONS

Relatively recently, removal of toxic heavy metal contaminants from industrial wastewater by biosorption has been proposed as a safe and cost-effective process for treatment of high volumes of low-concentration metal-polluted solutions. Biosorption could be combined with metal recovery by a desorption process [1,2]. Volesky and Holan [3] compiled a list of biomass and metals that have been tested for biosorption. Research on biosorption is revealing that it is sometimes a complex phenomenon where the metallic species are deposited in the solid biosorbent through different sequestering processes of chelation, complexation, and ion exchange. Muzzarelli [4] considered that chelation was the main mechanism for cupric ions binding to chitosan membranes. Complexation of metal by various heterogeneous complexants was explained well by Buffle and Altmann [5]. Darnall et al. [6], Kuyucak and Volesky [7], and Sharma and Forster [8] attributed the biosorption of heavy metals to chemical sorption. Treen-Sears et al. [9] presented evidence of ion exchange taking place upon biosorption of uranyl ion on *Rhizopus* biosorbent. Crist et al. [10,11,12] and Schiewer et al. [13] confirmed that ion exchange played an important role in biosorption of heavy metals on algal biomass. Since a large portion of heavy metal pollution may occur in the form of cations, a good proportion of biosorption research has been directed toward the removal of heavy metal cations.

The brown alga *Sargassum fluitans* (seaweed) has been found particularly effective in binding heavy metal ions of gold, cadmium, copper, and zinc [3]. The metal biosorption capacity of some more common (nonliving) marine algae was explored by Kuyucak and Volesky [14] and more recently by Figueira et al. [15]. The high sorption capacity, easy regeneration, and low costs make this biomass of special interest for the purification of high volumes of wastewater with lower concentration levels of metal toxicity to be removed [3,16]. This is difficult and/or expensive to accomplish by conventional metal-removal processes.

The most systematically studied has been the *Sargassum*-based metal biosorption system. Exciting high uptakes of uranium, a very threatening heavy metal, spurred more detailed investigations. While uranium mill tailings are a source of low-level radiation, 0.1–1.0 mR/h, uranium is very toxic. Various nonliving biomass types—including those of filamentous fungi, such as *Mucor meihei* [17], *Aspergillus niger* [18], *Rhizopus oryzae*, and *Penicillium* species [19,20]; bacteria such as *Pseudomonas aeruginosa*, *Zoogloea ramigera* [21,22], and *Citrobacter* [23]; actinomycetes such as *Streptomyces longwoodensis* [24]; yeasts such as *Saccharomyces cerevisea* [25,27]—have been reported to bind uranium in excess of 150 mg/g of dry biomass. Fresh water algae such as *Chlorella regularis* and *vulgaris* also demonstrated a good uranium adsorption performance [28,29]. Although the fact that live marine algae are capable of biologically concentrating radionuclides such as radium, thorium, and uranium has been known for a long time [30], the biosorption of uranium by nonliving marine algal biomass has rarely been reported.

There are several major factors in the biosorption process design and optimization: (1) equilibrium isotherm relationship, (2) biosorption process rate, and (3) breakthrough time in a continuous flow biosorption column.

A. Biosorption Equilibrium

In most cases, biosorption performance has been evaluated through the use of simple sorption isotherm relationships such as Langmuir or Freundlich models. They are defined, respectively, as follows.

Langmuir isotherm:

$$q = \frac{q_{max}[M]}{K + [M]} \tag{1}$$

where q and $[M]$ are equilibrium metal uptake and concentration, respectively, q_{max} is the maximum uptake, and K is the Langmuir sorption equilibrium constant.

Freundlich isotherm:

$$q = k[M]^{1/p} \tag{2}$$

where k and p are empirically determined constants: k is related to the maximum binding capacity and p to the affinity between sorbent and sorbate.

Both isotherms have been used successfully to describe biosorption equilibria [31–33]. However, it is worth noting that neither model can predict the effect of solution pH on the biosorption performance. Consequently, the determination of a series of isotherms at different solution pH values is always necessary. Considering ion exchange and binding to free sites, the Schiewer-Volesky model [34] incorporates the proton concentration in the sorption isotherm equation

$$q_{U} = \frac{0.5\,C_t\sqrt{K_x[\mathrm{X}]}}{1 + K_h[\mathrm{H}] + \sqrt{K_x[\mathrm{X}]}} \tag{3}$$

The prediction of the biosorption uptake at various solution pH levels then became an easy explicit calculation. However, the Schiewer-Volesky model neglected the possible hydrolysis of metal ions in the aqueous solution. For some metals, such as cadmium and zinc, simple metal ions are the predominant species in solution over a wide and relevant concentration and pH ranges. The effect of hydrolysis is then negligible. For some other metals, such as uranium, hydrolysis takes place under most biosorption experimental conditions. The existence of the hydrolyzed ions plays an important role in the biosorption of those metals. Guibal et al. [17] and Hu et al. [21] discussed qualitatively the effect of uranium hydrolysis on the biosorption by the filamentous fungus *Mucor meihei* and *Pseudomonas aeruginosa* strain CSU, respectively. It can easily be seen that it would be useful to develop a proper model equation that would be able to quantitatively describe the biosorption of different hydrolyzed metal species in solution, such as the case with uranium. This was achieved by the hydrolysed ion exchange model (HIEM), which was recently developed by Yang and Volesky [35]. The model, based on ion exchange, accounts for the hydrolysis of metal ions in the aqueous solution. The development of HIEM model equations will be demonstrated here for the case of uranium biosorption.

1. Uranium Biosorption Example

The hydrolysis equilibrium of uranium metal ions follows these stoichiometric relationships [36]:

$$2\ UO_2^{2+} + 4H_2O \leftrightarrow (UO_2)_2(OH)_2^{2+}$$
$$+ 2\ H_3O^+ \quad pK = 5.62 \tag{4}$$

$$UO_2^{2+} + 2H_2O \leftrightarrow UO_2OH^+ + H_3O^+ \quad pK = 5.8 \tag{5}$$

$$3UO_2^{2+} + 10H_2O \leftrightarrow (UO_2)_3(OH)_5^+$$
$$+ 5H_3O^+ \quad pK = 15.63 \tag{6}$$

where pK is the logarithm of the equilibrium constant.

In the diluted solution, the activity coefficient is close to unity and the hydrolysis equilibrium constants for Eqs. (4), (5), and (6) can be expressed in the following manner:

$$K_{ey} = \frac{[Y][H]^2}{[X]^2} = 10^{-5.62} \tag{7}$$

$$K_{ez} = \frac{[Z][H]}{[X]} = 10^{-5.8} \tag{8}$$

$$K_{et} = \frac{[T][H]^5}{[X]^3} = 10^{-15.63} \tag{9}$$

where K_{ey}, K_{ez}, and K_{et} are the equilibrium constants for Eqs. (4), (5), and (6), respectively. X, Y, Z, and T represent ionic species UO_2^{2+}, $(UO_2)_2(OH)_2^{2+}$, $(UO_2)(OH)^+$, and $(UO_2)_3(OH)_5^+$, respectively. As indicated by the very small value of K_{et} in Eq. (9), the effect of the uranium complex ion $(UO_2)_3(OH)_5^+$ in the aqueous uranium solution system on the biosorption of uranium can be neglected in the following model development.

The alginic acid polymers have been indicated as the main binding components in the *Sargassum* seaweed biomass. They are mainly arranged in flat ribbons with repeating units spaced at 1.04 nm. They are bound into sheets with hydrogen bonds [37]. The binding of divalent cations to the alginic acid has been considered to be in an egg carton pattern. The divalent cations are bound in a zigzag configuration between two alginic acid chains. For ion exchange between all forms of uranium ions and protons on the biomass binding sites, the following equilibrium equations have been proposed:

$$H^+ + B \leftrightarrow HB, \qquad K_h = \frac{[HB]}{[H][B]} \tag{10}$$

$$X + 2B \leftrightarrow 2X_{0.5}B \qquad K_x = \frac{[X_{0.5}B]^2}{[X][B]^2} \tag{11}$$

$$Y + 2B \leftrightarrow 2Y_{0.5}B \qquad K_y = \frac{[Y_{0.5}B]^2}{[Y][B]^2} \tag{12}$$

$$Z + B \leftrightarrow ZB \qquad K_z = \frac{[ZB]}{[Z][B]} \tag{13}$$

where K_h, K_x, K_y, and K_z are equilibrium formation constants corresponding to the bound components formed between various forms of uranium ions and protons with biomass in the solution. The formulation for the complex divalent uranium ionic groups and biomass binding sites is chosen as $2M_{0.5}B$ instead of M_2B. This is to emphasize that not only the electrostatic attraction but also complexation is relevant in this case where two bonds between the metal ion and biomass have to be broken in competitive binding or in desorption of the metal from biomass [38,39].

The total binding sites on the biomass are distributed among all the bound forms of hydrolyzed uranium ions and protons as well as the residual free sites:

$$C_t = [B] + [HB] + [X_{0.5}B] + [Y_{0.5}B] + [ZB] \tag{14}$$

where C_t and [B] are the concentrations of the total and free binding sites, respectively.

Upon substituting Eqs. (7)–(13) into Eq. (14), the concentration of free binding sites in the equilibrium system is obtained:

$$[B] = \frac{C_t}{1 + K_h[H] + \sqrt{K_x[X]} + \frac{[X]}{[H]}(\sqrt{K_{ey}K_y} + K_{ez}K_z)} \tag{15}$$

Substituting Eq. (15) into Eqs. (10)–(13), respectively, we obtain

$$q_H = [HB] = \frac{C_t K_h [H]}{1 + K_h[H] + \sqrt{K_x[X]} + \frac{[X]}{[H]}(\sqrt{K_{ey}K_y} + K_{ez}K_z)}$$

(16)

$$[X_{0.5}B] = \frac{C_t \sqrt{K_x[X]}}{1 + K_h[H] + \sqrt{K_x[X]} + \frac{[X]}{[H]}(\sqrt{K_{ey}K_y} + K_{ez}K_z)}$$ (17)

$$[Y_{0.5}B] = \frac{C_t \frac{[X]}{[H]}\sqrt{K_{ey}K_y}}{1 + K_h[H] + \sqrt{K_x[X]} + \frac{[X]}{[H]}(\sqrt{K_{ey}K_y} + K_{ez}K_z)}$$ (18)

$$[ZB] = \frac{C_t \frac{[X]}{[H]} K_{ez}K_z}{1 + K_h[H] + \sqrt{K_x[X]} + \frac{[X]}{[H]}(\sqrt{K_{ey}K_y} + K_{ez}K_z)}$$ (19)

Equation (16) could be used for the calculation of proton sorption on the biomass at any given uranium concentration level and solution pH. The total uranium uptake q_U is assessed as the summation of all the bound forms of uranium ions, i.e.,

$$q_U = 0.5[X_{0.5}B] + (2 \times 0.5)[Y_{0.5}B] + [ZB]$$ (20)

Upon substitution of $[X_{0.5}B]$, $[Y_{0.5}B]$, and $[ZB]$ with Eq. (17)–(19), the following equation was obtained:

$$q_U = \frac{C_t\left[0.5\sqrt{K_x[X]} + \frac{[X]}{[H]}(\sqrt{K_{ey}K_y} + K_{ez}K_z)\right]}{1 + K_h[H] + \sqrt{K_x[X]} + \frac{[X]}{[H]}(\sqrt{K_{ey}K_y} + K_{ez}K_z)}$$ (21)

In Eq. (21), the uranium uptake could be calculated from the solution pH or proton concentration, [H], and the concentration of the free uranium ion UO_2^{2+}, [X]. Although the mathematical form of Eq. (21) is similar to that for the multicomponent Langmuir sorption isotherm as adapted from Hill [40], the mechanism that lies behind it is completely different. The present model considers ion exchange, not only a simple competition for free binding sites where no reverse reaction takes place. In Eq. (21), the concentration of the free uranium ion [X] is not necessarily the same as the total uranium concentration in the solution because of the hydrolysis of the uranium ion in the solution. In other words, [X] could not be measured directly, and it must be evaluated from the measurable total uranium concentration through the hydrolysis equilibrium calculation. Usually, this could be done by using a computer program such as MINEQL + [41], computer software for the calculation of the chemical equilibrium composition of aqueous solutions. However, it would be more useful to derive a formula expressing the concentration of the complexion ions as an explicit function of the measurable total uranium concentration and the pH value.

For this purpose, the total uranium concentration U_t in the solution is considered to consist of free ions and all major hydrolyzed ions:

$$U_t = [UO_2^{2+}] + 2[(UO_2)_2(OH)_2^{2+}] \\ + [(UO_2)(OH)^+] = [X] + 2[Y] + [Z] \tag{22}$$

Substituting [Y] from Eq. (7) and [Z] from Eq. (8) into Eq. (22),

$$U_t = [X] + \frac{2K_{ey}[X]^2}{[H]^2} + \frac{K_{ez}[X]}{[H]} \tag{23}$$

The concentration of free uranyl ions $[UO_2^{2+}]$, or [X], in the solution can be obtained as the following equation:

$$[X] = \frac{[H][\sqrt{([H] + K_{ez})^2 + 8K_{ey}U_t} - ([H] + K_{ez})]}{4K_{ey}} \tag{24}$$

Equations (16), (21), and (24) express the proton uptake and the uranium biosorption uptake as an explicit function of the total uranium concentration and the solution pH value, which fulfils the main modeling objective.

2. Electroneutrality

When LiOH is added to the system to maintain a constant solution pH, the electroneutrality condition is given by

$$[NO_3^-] + [OH^-] = [H^+] + [Li^+] + 2[UO_2^{2+}]$$
$$+ 2 [(UO_2)_2(OH)_2^{2+}] + [(UO_2)(OH)^+] \tag{25}$$

where the nitrate is introduced as a co-ion in the initial uranyl nitrate solution, while the nitrate concentration $[NO_3^-]$ remains constant in the system. Based on water equilibrium, the concentration of hydroxyl groups can be expressed as

$$[OH^-] = \frac{10^{-14}}{[H^+]} \tag{26}$$

Substituting Eq. (22) and (26) into Eq. (25), the concentration of counterions maintaining the electroneutrality can be calculated from the equilibrium uranium concentration and proton concentration as follows:

$$[Li^+] = [NO_3^-] - [U_t] - [X] - [H^+] + \frac{10^{-14}}{[H^+]} \tag{27}$$

where [X] can be obtained from Eq. (24).

It should be noted that the charge balance does not affect the validity of the mass balance on the binding sites, as the diffusion of any heavy metal ion into the biomass would cause the diffusion out of the equivalent protons from the biomass. The anions (nitrate) do not associate with any particular cations, but they accompany the transfer of cations into and out of biomass. If the hydrolysis could be neglected, the hydrolysis constants in Eqs. (7) and (8) would be zeros, i.e., $K_{ey} = K_{ez} = 0$. The UO_2^{2+} concentration [X] would be the same as the total uranium concentration U_t in Eq. (22). Then the model Eq. (21) would be reduced to the Schiewer-Volesky [34] model equation for the nonhydrolyzed metal ions such as, for example, those of cadmium, copper, or zinc.

The hydrolysis equilibrium constants for uranium ions UO_2^{2+}, K_{ey}, and K_{ez} in Eqs. (7) and (8) are thermodynamic constants. The total biomass binding capacity C_t can be determined by acid–base titration of the *Sargassum* biomass [39,42]. The four biosorption equilibrium constants, K_h, K_x, K_y, and K_z, for protons and hydrolyzed uranium complexion ions UO_2^{2+}, $(UO_2)_2(OH)_2^{2+}$, and UO_2OH^+ can be regressed from the uranium isotherm experimental data at various pH values. The experimental uranium isotherm data and the model regressed curves at different pH

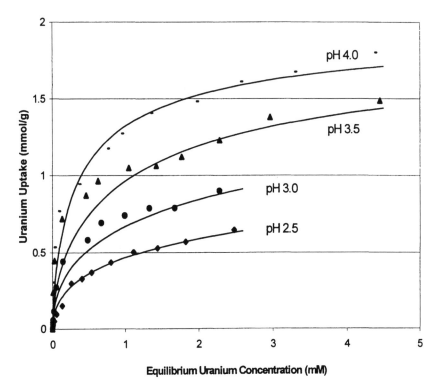

Figure 1 Comparison of experimental uranium isotherms and HIEM calculation at different solution pH.

(2.5–4.0) are illustrated in Fig. 1. The corresponding model parameters are listed in Table 1.

If the hydrolysis of uranyl cation UO_2^{2+} were not taken into consideration, the maximum uranium uptake by ion exchange between U and H^+ would be half of the biomass binding capacity, 1.125 mmol/g, which

Table 1 HIEM Model Parameters for Uranium

C_t (mmol/g)	K_h (L/mol)	K_x (L/mol)	K_y (L/mol)	K_z (L/mol)
2.25	235.1	1081.7	1.7731×10^4	1.1494×10^4

is much less than the maximum experimental uranium uptake, 1.78 mmol/g, observed at pH 4.0, as seen in Fig. 1.

Furthermore, since the proton concentration is included in the HIEM model, the model-calculated uranium uptake is a function of both independent variables, pH and the equilibrium uranium concentration. In this situation, an isotherm surface is more appropriate for depicting the biosorption performance, as demonstrated in Fig. 2.

The scattered points represent the experimental uranium uptakes and the three-dimensional surface was plotted from model-calculated uptake values. With increasing solution pH and particularly uranium concentration, the uptake surface becomes higher. In order to demonstrate the effect of pH on the uptake at different equilibrium uranium concen-

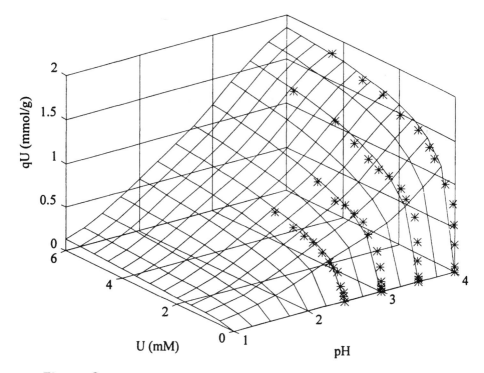

Figure 2 Uranium sorption isotherm experimental data and HIEM model regression at solution pH 2.5–4.0 and uranium metal concentrations 0.0–6.0 mmol/L. *, experimental; mesh, HIEM model.

trations, a series of curves were obtained by cutting the sorption isotherm surface at given uranium equilibrium concentrations. These relationships are plotted in Fig. 3. For different uranium concentrations, the uranium uptake increases with increasing solution pH. The effect of solution pH on the metal uptake appears less significant at lower uranium concentrations compared to that at higher uranium concentration.

It has been well established that sorption of many types of metal species, such as Cd^{2+}, Cu^{2+}, Zn^{2+}, Pb^{2+}, Ni^{2+}, Mn^{2+}, Al^{3+}, and Co^{2+}, increases with increasing solution pH as the metal ionic species become less stable in the solution [7,43–48]. Only a few metal ions, such as Ag, Au, and Hg, in aqueous solution either form negatively charged complexes or have strong tendency to sorb by forming covalent bonds. In those cases, the increase in solution pH would decrease the sorption uptake or would not affect the sorption at all [46,47,49].

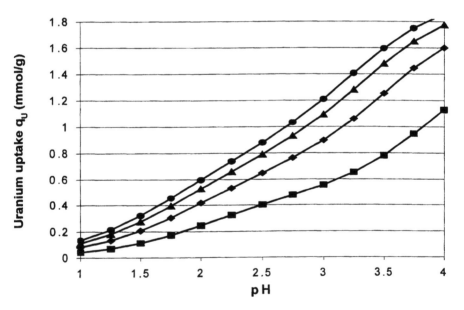

Figure 3   Influence of solution pH on the uranium sorption at fixed uranium equilibrium concentrations. ●, $U_f = 6.0$ mmol/g; ▲, $U_f = 4.0$ mmol/g; ◆, $U_f = 2.0$ mmol/g; ■, $U_f = 0.5$ mmol/g.

The ionic distribution in the biomass phase, calculated by using model Eqs. (17), (18), and (19) together with Eq. (24), is presented in Fig. 4. The $(UO_2)_2(OH)_2^{2+}$ already exists even at very low pH values. For instance, there is more than 20% of bound uranium in the form of $(UO_2)_2(OH)_2^{2+}$ at pH 2.5 if the total uranium concentration is high enough. Whereas the corresponding total equilibrium uranium concentration had a minor effect on the ionic distribution in the biomass, the percentage of $(UO_2)_2(OH)_2^{2+}$ in the biomass phase increases significantly as the corresponding solution pH level in the biosorption equilibrium increases. At pH 4.0, approximately 70–80% of total uranium exists in the $(UO_2)_2(OH)_2^{2+}$ form. In addition, although in a relatively lower amount,

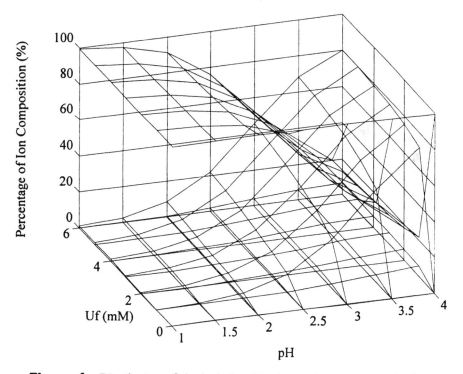

Figure 4 Distribution of the hydrolyzed ionic uranium species in the biomass. Top mesh, UO_2^{2+}; middle mesh, $(UO_2^{2+})_2(OH)_2^{2+}$; bottom mesh, $(UO_2^{2+})(OH)^+$.

UO_2OH^+ ions exist over a wide range of pH values. Since the complex ion contains twice the number of uranium atoms than the UO_2^{2+} ion for the same electrical charge; the existence of those complex ions in the biomass phase indicates that the overall uranium uptake must be higher. This conclusion was well supported by the experimental data as described in the previous sections.

After the biosorption of toxic heavy metals from large volumes of low-concentration aqueous waste, the sequestered heavy metal could be eluted from the saturated biomass by a desorption or an elution process. Aldor et al. [2] investigated equilibrium cadmium desorption from protonated *Sargassum fluitans* biomass by various elutants. It was established that the mineral acids, particularly 0.1 N HCl or H_2SO_4, are efficient in metal elution,and the biomass damage was limited during the acid wash.

B. Biosorption Batch Dynamics

For any practical applications, process design, and operation control, the sorption process rate and the dynamic behavior of the system are very important factors. Generally, biosorption of metal ions consists of three continuous processes. The metal ions first diffuse across the particle-to-fluid film from the bulk solution before they enter the biosorbent. Then they diffuse further toward the binding sites through the gel phase of the biomass material. Finally, the metal ions react with the functional groups of the binding sites. External and internal mass transfer resistance by the biomass particle are of concern for the sorption dynamics. The intraparticle mass transfer has been established as the rate controlling step by many researchers. In order to determine the intraparticle mass transfer rate experimentally, the effect of external mass transfer must be eliminated first. The external mass transfer resistance is proportional to the thickness of the stationary fluid layer or film surrounding the biomass particles, which, in turn, is controlled by the agitation in the bulk solution. Strong stirring will decrease the film thickness and could eventually eliminate the film resistance if the stirring is strong enough. A series of sorption experiments were carried out with increasing agitation rates until the produced concentration profiles were identical within a acceptable error range.

In order to determine the biosorption rate at a constant solution pH, the protons released from *Sargassum* biomass to the solution as the result of the biosorption of metal cations must be neutralized instantly. This can be achieved by using an autotitrator assembly in an end point titration mode. When the solution pH decreased below the predefined pH value (end point) because of the proton release, the internal burette of the autotitrator was activated and delivered LiOH solution into the reactor instantly until the solution pH was restored to the designed level. Biosorption of metal ions took place very rapidly during the initial stage of contact. The concentration profile curve for uranium is shown in Fig. 5.

The sorption rate appeared to go through two stages. The first stage was fast and about 75% of the total cadmium adsorption took place in 15 min. The second stage was relatively slower and it took approximately 3 h to reach the sorption equilibrium. Applying the first-order rate method

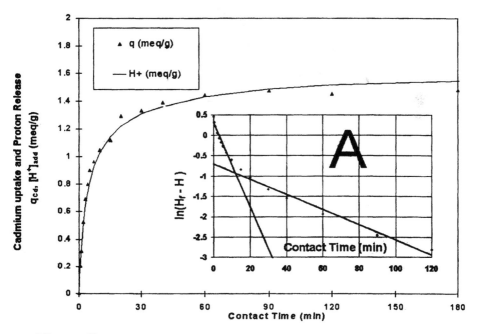

Figure 5 Uranium biosorption rate and proton release rate at pH 4.0. A, two types of uranium biosorption rates.

introduced by Crist et al. [50] for the exchange of calcium ions for protons on a peat moss biosorbent, the sorption rate was assumed proportional to the binding capacity of the biomass, i.e.,

$$d[H]/dt = k([H]_\infty - [H]) \tag{28}$$

where t is the contact time, k is the rate constant, and $[H]$ and $[H]_\infty$ stand for the instantaneous and maximum proton release, respectively. The integrated form of the first-order rate equation gave

$$\ln([H]_\infty - [H]) = kt \tag{29}$$

The plots of simulations using Eq. (29) with experimental data for uranium are shown in Fig. 5A. The slope of the lines represents the sorption rate constant. The slope for the initial 10 min was calculated as k_1 = −0.123 and that for the remaining contact time was k_2 = −0.0214. The difference in the slope values is significant; k_1 is approximately 5.7 times k_2. This reflects the two different process rates occurring in the initial stage and the last stage more clearly. This type of a rate curve has often been reported by other researchers. Beside the Crist et. al. [50] explanations, Brassard et al. [51] attributed the fast initial metal sorption rate to surface binding on natural particles and the following slower sorption to interior penetration. In the case of *Sargassum* seaweed particles, the active binding groups reside in the cell wall, and due to its large surface the initial sorption rate is accelerated. The actual mechanism of heavy metal sequestration has been studied only to a limited extent [10] and should be studied further.

It was physically difficult to sample the reaction solution at the early stage of the sorption process, which has a significant effect on the overall sorption rate. This was a serious limitation to the accuracy of the data regression process used for determining the diffusion coefficient. The agreement between the metal uptake and the proton release during the sorption process indicates an equivalent ion exchange between uranium or cadmium ions and protons. With a fast pH probe response and titration speed, the computer-recorded titration volume of the added alkali LiOH versus contact time (up to 8 points per second) curves could be used to regress the effective diffusion coefficient directly. This method provides a useful alternative to the evaluation of the sorption rate, avoiding the tedious procedure of reactor solution sampling for metal concentration determination.

A mathematical model is a very desirable tool delivering a fair representation of behavior of all important metals that can be sorbed by the kind of biosorbent material considered. In order to do so the model has to incorporate the dynamic process parameters including the rate control aspects.

For a quantitative description of the biosorption process dynamics it is necessary to develop a mathematical model capable of reflecting the toxic metal ion concentration change with contact time. Under the experimental conditions, the particle-to-liquid mass transfer resistance has been eliminated through adequate turbulence created by proper agitation. The intraparticle diffusion of cadmium ions is assumed as the rate controlling step, the more detailed model assumptions are made as follows:

1. The sorption reaction between the metal ion and binding sites on the biomass is much faster than diffusion of the metal cation inside the biomass material, i.e., the overall sorption rate is controlled by the intraparticle diffusion.

2. For the relatively flat seaweed *Sargassum* biomass particle, the thickness of the particle is much smaller than the length and width. Thus the biomass particle can be considered as a thin plate. The intraparticle diffusion along one dimension, i.e., the thickness direction, controls the diffusion rate of the overall process.

3. The amount of adsorbed metal ions inside the biomass particle is in equilibrium with the metal concentration in the liquid phase and the Langmuir sorption isotherm relationship of Eq. (1) holds.

It must be indicated that the Langmuir adsorption assumption did not reflect the experimental results that demonstrated a protons release as a result of ion exchange with the metal ions. Thus the diffusion of metal ions into the biomass must accompany the protons diffusing out to the bulk liquid. However, the diffusion coefficient of H^+ is several times higher than that of heavy metal ions in infinitely diluted aqueous solutions [52,53]. In addition, Crist et al. [54] show that the Langmuir model can be used for ion exchange rate calculations for algal biomass over a wide concentration range. Thus it is reasonable to assume that the overall sorption rate was solely controlled by the heavy metal ions.

Based on the model assumptions, the mass conservation equations for metal ions in the biomass particle and bulk solution are as follows:

$$\varepsilon \frac{\partial C_r}{\partial t} = \rho \frac{\partial q}{\partial t} = D_e \frac{\partial^2 C_r}{\partial r^2} \tag{30}$$

$$V \frac{dC_b}{dt} = D_e S_t \frac{\partial C_r}{\partial r}\bigg|_{r=R} \tag{31}$$

where r is the arbitrary position coordinate from the central line of the biomass particle in the thickness direction and t is the time elapsed from the start of the sorption process; C_b and C_r represent the metal concentrations in the bulk solution and in the pore liquid phase at layer r inside the biomass, respectively; R is the half-thickness of the biomass particle, ε is the porosity of the biomass, and ρ is the density of the biomass; V is the volume of bulk solution and S_t is the total surface area of particles that can be calculated from the geometry of the particle assumption; and D_e represents the effective intraparticle diffusion coefficient.

The boundary and initial conditions for the sorption process are as follows:

$$C_{r|_{r=R}} = C_b \qquad (r = R, \ t > 0) \tag{32}$$

$$\frac{\partial C_r}{\partial r}\bigg|_{r=0} = 0 \qquad (r = 0, \ t > 0) \tag{33}$$

$$C_b = C_0 \qquad (t = 0) \tag{34}$$

$$C_r = 0 \qquad (t = 0, \ 0 \leq r < R) \tag{35}$$

where C_0 is the initial metal concentration in the solution.

Differentiating the isotherm Eq. (1), we obtain

$$\frac{\partial q}{\partial C_r} = \frac{K q_m}{(K + C_r)^2} \tag{36}$$

Equations (30) and (31) are simultaneous nonlinear partial differential equations (PDEs) with respect to C_r and C_b, respectively. They are similar to the ones used in diffusion of a gas in microvoids of a polymer. For a linear isotherm, an analytical solution was presented by Crank [110].

In the case of nonlinear isotherms, such as the Langmuir isotherm used in Eq. (1), the analytical solution is not available so that a numerical method must be introduced to solve the complicated equations.

There are many numerical methods that can be used to solve PDEs. Among them, the Galerkin finite element method (GFEM) [55] and orthogonal collocation (OC) are frequently used. Once the model PDEs are discretized to a series of ordinary differential equations (ODEs), the Euler backward integration or Runge-Kutta integration in time can be applied to obtain the numerical solution for the intraparticle concentration C_r and concentration C_b.

Among the model parameters, K and q_m can be obtained from batch equilibrium experiments. R, ε, and ρ can be measured directly and S_t (in cm^2) is decided by the weight of the biomass used, W, for the thin plate geometry of the biomass particle; $S_t = W/\rho R$. All other parameters, except D_e, can be calculated once the volume V and the concentration of the metal solution in the reactor are specified. This means that the numerical solution of the model equations is uniquely determined by the value of D_e. In other words, a specific value of D_e corresponds to a specific simulated concentration profile of the $C(t)$ versus t curve. The intraparticle diffusion coefficient D_e can thus be regressed from the comparison of the simulated profile curves and the experimental results.

An example of simulation is conducted for uranium biosorption rate at pH 2.5–4.0. The corresponding D_e values were regressed in Table 2. The average values of diffusion coefficients are 5.5×10^{-6} cm^2/s for uranium. All values of the determined diffusion coefficients are in the same order of magnitude as their respective molecular diffusivity, 7.21×10^{-6} cm^2/s for uranium [52,53]. In the literature, the values of the effective diffusion coefficient are very diversified and can differ by several orders of magnitude even at similar experimental conditions. The comparison of model-fitted concentration profiles and the experimental data points at pH 2.5–4.0 are plotted in Fig. 6.

Table 2 The Regressed Diffusion Coefficients D_e (cm^2/s)

	pH 2.5	pH 3.0	pH 3.5	pH 4.0	Average
Uranium	4.0	6.0	6.0	6.0	5.5

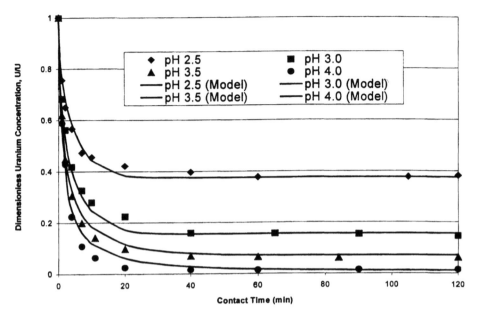

Figure 6 Modeling uranium concentration–time profiles at different solution pH values.

In general, the diffusion process inside a porous material is slower than that in a corresponding homogeneous system having the same liquid composition as in the pore phase [56,57]. In the brown alga *Sargassum* biomass, alginate is the main component responsible for the metal sorption [58]. It is present in a gel form in the cell wall, which appears very porous and easily permeable to small ionic species [59,60]. The actual mobility of the diffusing entity in the dense-phase gel may be somewhat reduced by mechanical friction or interaction with the cell wall molecules. As a result, the calculated intraparticle diffusion coefficient is an effective diffusion coefficient, D_e, and it is usually smaller than the molecular diffusion coefficient D_m considered in the absence of the sorbent material matrix.

C. Dynamics of Biosorption in a Continuous System

Most practical and effective applications of adsorption for separation and purification in industry have commonly been carried out in fixed-bed columns. In the column operation system, the feed percolates continuously through a column filled with the sorbent that retains the sorbate component(s) from it. The sorbent material in the column becomes gradually saturated from the feed zone to the exit as the adsorption process progresses. When the concentration of the sorbate reaches a predefined level in the exit stream (the "breakthrough"), the column operation is stopped and the regeneration process often takes place as the next stage of operation.

The breakthrough concentration versus time profile is the most important characteristic of a column system. During laboratory uranium biosorption column operation, the acidic influent with the uranium concentration of $C_0 = 238$ ppm (pH 2.5) passed through the column upward at a flow rate of $F = 340$ mL/h. The empty bed volume ($V_{bed} = 280$ cm^3) of the column was filled with $W = 22.64$ g (dry) protonated *Sargassum* seaweed biomass. The breakthrough curves are shown in Fig. 7, where the uranium concentration and pH value in the column outlet are plotted versus the dimensionless volume (the column volume being the reference) of uranium-containing fluid that passed through the column. The ura-

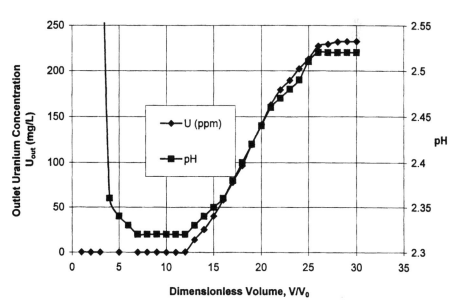

Figure 7 Uranium biosorption column operation: the breakthrough curve.

nium breakthrough curve can be seen as a typical "S" shaped break-through curve. The breakthrough took place at 36.5 bed volumes, when 1.0 ppm uranium was detected in the effluent at the column outlet. Before the breakthrough time, approximately 10 L of 238 ppm uranium solution was processed. In this experiment, the average hydraulic residence time was about 49 min and a total amount of 2380 mg uranium was accumu-lated on 22.64 g (dry) of biomass, giving the average column uranium biosorption capacity of 105 mg U/g (dry biomass). When the column started, the initial effluent pH represented the pH level of prerinsing dis-tilled water. Eventually, the effluent pH decreased and its variation matched the uranium exit concentration profile, demonstrating that ion exchange between uranium metal ions and protons took place in the bio-sorption zone. This behavior could conveniently even be used to approxi-mately determine the uranium breakthrough point with ease since the detection of pH is much simpler than determination of the uranium con-centration.

The uranium-saturated column was desorbed (regenerated) by pumping 0.1 N HCl acid through it at the same flow rate as was used earlier for the uranium biosorption. The elution results are shown in Fig. 8, where the uranium concentration in the elutant acid (0.1 N HCl) at the outlet of the column and the effluent pH were plotted against the volume of the elutant acid passed through the column.

The outlet uranium concentration in the desorbing solution showed a narrow peak of about 6000 mg/L average concentration in the 400 mL of acid volume. The overall performance of the biosorption process is indicated by the overall concentration factor defined as the ratio of the elution concentration to the influent concentration in the sorption run, i.e., $CF = U_E / U_0$. In this elution run, the concentration factor was deter-mined as $CF_U = 25$, and more than 99.5% of sorbed uranium was recov-ered from the column. The narrow elution peak and the low residual uranium concentration (<1 ppm) indicated that the elution was highly efficient. Furthermore, when a lower elution flowrate, e.g., 175 mL/h, was employed, an even higher uranium wash-peak concentration of 11,000 mg/L was obtained. During the acidic elution process, the biosor-bent was prepared for another sorption cycle as its metal-binding sites became protonated.

With simple distilled water rinses removing residual acid, the col-

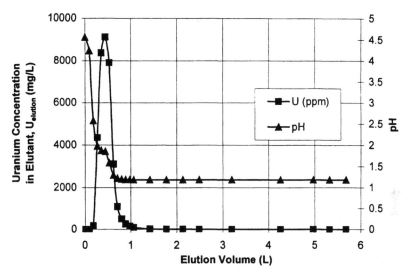

Figure 8   Elution of uranium with 0.1 N HCl from biosorption column.

umn was operated in five cycles for a whole month. The breakthrough curves for different cycles are illustrated in Fig. 9. Although the shape of breakthrough curves remained almost unchanged, the breakthrough time decreased slightly mainly only after the first cycle, indicating that the biomass binding capacity in the column was slightly diminished. The biosorption capacity in the fifth cycle was about 20% lower than that for fresh biosorbent, but only about 7% lower than that in the second cycle. Figueira et al. [15] correlated the metal-binding loss to the amount of TOC (total organic carbon) detected in the effluent during a biosorption column operation. However, no significant visible damage to *Sargassum* biosorbent was observed in one month of continuous operation.

1.   Modeling the Column Performance

The design of column systems requires knowledge of the time for which the adsorbent bed can be exposed to the fluid before breakthrough of the sorbate occurs. The breakthrough time will be determined by the equilibrium isotherms of the adsorbed metal components, the fluid flow, the

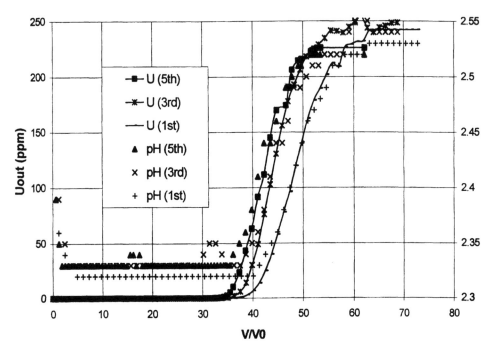

Figure 9 Breakthrough curves for different sorption–desorption cycles. $F =$ 340 mL/h; $V_{bed} = 280$ cm^3, biomass $= 22.642$ g.

axial and radial dispersion of the fluid in the column, and the external and internal diffusion resistance for the sorbate material. Although the most important mass transfer resistance for adsorption from aqueous solutions, such as wastewater, has been established to be the internal (intraparticle) diffusion by many researchers [61,62], film resistance may also play a role [63]. In order to apply the scale-up calculation in a wide range of cases, it is important to represent the dynamic column behavior by a proper mathematical model that can be solved either analytically or numerically.

A methodology similar to the one described for batch systems can be used to develop a model for continuous flow column systems. Crittenden, Weber and Asce [64–66] established that column flow dispersion effects could generally be neglected for relatively long columns. Liu and Weber's [67] studies of microcolumn systems show that the effect of the axial dispersion was negligible as well. The axial dispersion effects were thus not included in the present mass transfer model for the biosorption column systems. The model proposed here is based on the same assumptions as described for batch dynamics.

The mass conservation equation in the macroscopic fluid is represented by

$$\varepsilon \frac{\delta C_b}{\delta \tau} + U_s \frac{\delta C_b}{\delta z} + \rho(1 - \varepsilon)\frac{\delta \bar{q}}{\delta \tau} = 0 \tag{37}$$

Film mass transfer:

$$\frac{\delta \bar{q}}{\delta \tau} = K_f a(C_b - C_r|_{r=R}) \tag{38}$$

Intraparticle diffusion:

$$\varepsilon_p \frac{\delta C_r}{\delta \bar{\tau}} + \rho \frac{\delta q}{\delta \bar{\tau}} = D_e \frac{\delta C_r}{\delta r^2} \tag{39}$$

Isotherm:

$$q = f(C_r) \tag{40}$$

C_b and C_r represent the metal concentration in the bulk fluid stream and in the fluid in the pore of the biosorbent, respectively, and q is the metal uptake on the solid phase-, τ and $\bar{\tau}$ represent the time related to the bulk fluid and the intraparticle diffusion, respectively; ε and ε_p are, respectively, the bed porosity and intraparticle porosities; ρ is the pellet density and U_s is the fluid superficial velocity, $U_s = F/S$, where F is the feed flow rate and S is the section area of the column; K_f is the mass transfer coefficient across the fluid-particle interface, and a is the specific external particle area, i.e., the total external surface area per unit volume of the particle; D_e is the effective diffusion coefficient inside the particle; and $f(C)$ is the sorption isotherm relationship.

The initial and boundary conditions for Eq. (37) and (39) are as follows:

$$\tau \leq 0, \ C_b = 0 \qquad\qquad (0 \leq z \leq L) \tag{41}$$

$$z = 0, \ C_b = C_{in} \qquad\qquad (\tau > 0) \tag{42}$$

$$\bar{\tau} \leq 0, \ C_r = 0 \qquad\qquad (0 \leq r \leq R) \qquad (43)$$

$$r = 0, \ \left. \frac{\delta C_r}{\delta r} \right|_{r=0} = 0 \qquad\qquad (\bar{\tau} > 0) \qquad (44)$$

$$r = R, \ D_e \left. \frac{\delta C_r}{\delta r} \right|_{r=R} = K_f(C_b - C_r|_{r=R}) \qquad (\bar{\tau} > 0) \qquad (45)$$

Since the axial dispersion was neglected, the biosorbent in a different axial position in the column is exposed to the fluid at a different time. The following relationship holds:

$$\bar{\tau} = \tau - \frac{\varepsilon z}{U_s} \qquad (46)$$

The dimensionless PDE can be solved numerically by the orthogonal collocation method as suggested by Villadsen and Michelsen [68]. The intraparticle diffusion coefficient D_e was determined by batch dynamics. The film coefficient K_f can be regressed from the experimental breakthrough curves. A trial and error procedure was adapted by adjusting the value of K_f until the regression objective function reached its minimum value.

Figure 10 illustrates experimental and model-calculated breakthrough curves for uranium ions sorbing to the protonated biomass. In the column operation, 22.6 g (dry) of biomass was loaded in the 280 cm³ bed volume. The pH 2.5 and 1.0 mM concentration of uranium solution was fed into the column at a flow rate of 340 mL/h. The isotherm parameters at pH 2.5 were adapted from the Langmuir regression, and the biomass particle properties are from the batch dynamics system as above. The value of the effective intraparticle diffusion coefficient 6.0×10^{-6} cm²/s, obtained for the batch system, was applied in this model. The bed porosity was determined by Kratochvil [69] as 0.77. The value of the external mass transfer coefficient K_f was therefore regressed as 3.0×10^{-3} cm/s. The fitting curve was close to the experimental one within an average deviation of 5%. Although K_f is correlated with hydraulic conditions such as Reynolds number, kinematics viscosity of the fluid [70–73], it is more convenient to numerically regress it from column experimental data. Weber and Liu [74] obtained a range of values for $K_f = 2.0–5.0 \times 10^{-3}$ cm/s using a microcolumn technique. The model-regressed value of $K_f = 3.0 \times 10^{-3}$ cm/s obtained by us agrees well with Weber and Liu's

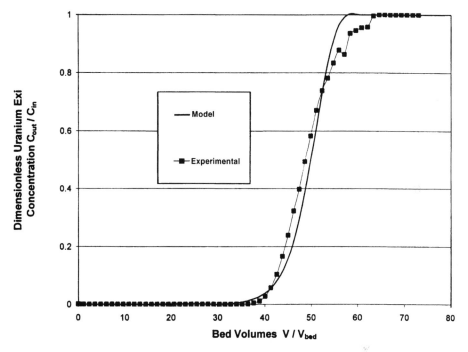

Figure 10 Comparison of the experimental uranium breakthrough curve and the mass transfer model calculated breakthrough curve for the protonated *Sargassum* biomass.

result [74]. However, it must be indicated that the model curve was not very sensitive to the external mass transfer diffusion coefficient. This implied that the external mass transfer in the biosorption system may not play an important role as compared with the intraparticle diffusion coefficient. The model-calculated breakthrough curve included the information from the biosorption equilibrium isotherm, the batch biosorption rate, and the fluid flow in the fixed-bed column.

D. Cation Biosorption Modeling Conclusions

The model simulation results can be applied in the biosorption process and scaling-up design, as well as in the optimization of the biosorption column operation. It must be indicated that the loading of the seaweed biosorbent into a large-scale column may be different from that used for laboratory columns. Many factors such as channeling and wall effects may lead to some problems usually encountered in attempts to apply mathematical models for estimation of large-scale column performance. A series of pilot experiments may be required to better verify and/or modify the mathematical model so that it can eventually be reliably used to guide scaling-up and design procedures.

III. BIOSORPTION OF ANIONS

While a large portion of the current biosorption research has been directed toward the removal of heavy metal cations, the uptake of anions by biomass has remained a challenge in the field of biosorption. The removal of molybdate (MoO_4^{2-}) by chitosan beads has recently been studied [75]. Removal of hexavalent chromium by peat moss [76], corncob [77], and other types of inexpensive waste biomass [8] has been reported. Chromium(VI) anion biosorption by *Sargassum* biomass was explained by a combined ion exchange and reduction mechanism [78]. An earlier patent has been granted for the use of *Sargassum* biomass in extracting gold from its chloride solutions [79,80]. Although still only little investigated, recent research into biosorption of anionic metallic species revealed a significant potential for using biomaterials in the anion removal/recovery process.

A. Biosorption of Gold Cyanide Anionic Complex

The recent studies on biosorption of the anionic $AuCN_2^-$ complex could well serve for the illustrative purpose of this chapter as an example of biosorption performance characterization using several typical biosorbent materials, namely bacterial (*Bacillus*), fungal (*Penicillium*), and algal (*Sargassum* seaweed) biomass types [81,82]. The $AuCN_2^-$ complex occurs in the conventional cyanide leach solution resulting in most commercial gold ore extraction processes [83]. As $AuCN_2^-$ is a very stable anion complex [84], the characteristics of $AuCN_2^-$ biosorption could conveniently be studied while the complexity of chemical speciation in the solution could reasonably be neglected. Needless to say, the principles and biosorption characteristics discussed in this section could easily be extended to similar types of biosorption behavior involving other anionic metal species including metals such as arsenic, selenium, vanadium, and others of interest occurring in solutions as anionic forms.

There are several major parameters that affect the biosorption of anions from solution—mainly the pH, ionic strength, and interference of other anions present in the sorption system, some of which will be briefly discussed in the following paragraphs. The focus will be on the biosorption performance of three above-mentioned representative types of biomass (*Bacillus, Penicillium,* and *Sargassum*).

1. pH Effect

Biosorption of the anionic $AuCN_2^-$ complex by any of the protonated *Bacillus, Penicillium,* or *Sargassum* biomass was strongly affected by pH [81]. Equilibrium uptakes of Au increased as the solution pH decreased. Figure 11 shows the Au biosorption isotherms obtained with, respectively, protonated *Bacillus, Penicillium,* and *Sargassum* biomass at equilibrium pH values of 2–6. At pH 2, the Au uptakes were higher than those at pH > 2 through the entire range of Au final concentrations from 0 to 20 mg/L. This concentration range was examined because it corresponds to that typical for conventional gold cyanide leach solutions [83]. At pH 2, *Bacillus* biomass could accumulate Au up to 8.0 μmol/g biomass, *Penicilium* biomass retained 7.2 μmol/g, and *Sargassum* biomass only 3.2 μmol/g. Negligible uptake of Au was obtained at pH 5.0, which was similar to the *Aspergillus* biomass performance [85].

The sorption system pH had a tendency to increase during the equilibration, requiring 0.1 M HNO_3 additions for pH adjustment. The same behavior was observed with anionic Cr(VI) adsorption by *Sargassum* biomass [78]. Lower pH values also yielded increased uptakes of anionic Cr(VI). Kratochvil et al. [78] described *Sargassum* biomass as an anion exchanger removing Cr(VI) from aqueous solution by the "acid adsorption" that was established as a mechanism of ion exchange on weak-base anion exchange resins [biomass–(H^+)–anion]. Basically, dead biomass material in aqueous solution has surface charges on ionizable functional groups. The charge due to a proton is one of the main important solid/liquid interface chemical features [86]. The surface charge is dependent

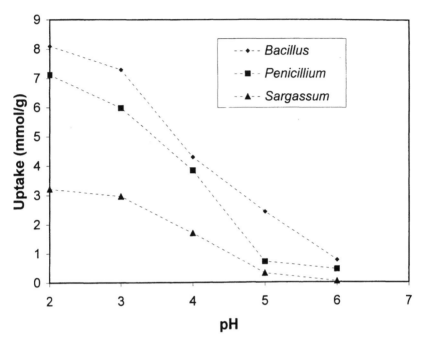

Figure 11   Effect of pH on Au biosorption by ◆, *Bacillus subtilis*; ■, *Penicillium chrysogenum*; ▲, *Sargassum fluitans*. Equilibrium contact of 0.04 g biomass, 20 mL solution, 4h, room temperature. (From Ref. 81.)

on the types of compounds in cell walls of the biomass mainly responsible for biosorption [3]. Cell walls of gram-positive bacteria, e.g., *Bacillus*, are mainly composed of peptidoglycan, which accounts for approximately 40–90% of the cell wall material [87]. Proteins and amino acid residues are also found in these cell walls. The pK of carboxyl groups in the above cell wall components is around 4.8 [88], and amine groups have a pK around 7–10 [89]. As the pH of the solution decreased from 6 to 2, negatively charged carboxyl groups and neutral weak-base amine groups on biomass became protonated, offering positive binding sites [biomass–(H^+)] for anionic gold cyanide complexes, which also became less repelled by the decreasing overall negative biomass charge.

    Similarly, as fungal cell walls contain up to 40% of chitin, often complexed with proteins, lipids, glucan, and other substances [90], they contain an abundance of amine and carboxyl groups as well. While in brown algae, cell wall alginate could make up to 45% of the cell wall dry weight [91], its active carboxyl groups have pKa of about 3–5 [89]. Thus anion adsorption by *Sargassum* biomass could rather be ascribed to phenolic groups of polyphenolic compounds such as polyphlorogluclnols, which are known to be present in brown seaweeds [59]. The pK of the phenolic OH is around 10 [89]. While the inorganic hydroxide could take up an extra proton and serve as a weak base for anion adsorption [92], the capacity of phenol OH in biomolecules to take up an extra proton at acidic pH is questionable. This is probably the reason why *Sargassum* biomass showed a much lower capacity for binding the gold cyanide complex than *Bacillus* or *Penicillium* biomass, which contains typically weak-base amine groups.

The isoelectric point of most microbial and brown algae biomass is in the range of pH 2–3.5 [93,94]. The occurrence of positively charged sites available for anionic $AuCN_2^-$ adsorption is not only determined by the solution pH and the charge state of biomass, but also by the number of weak-base groups on biomass that could become positively charged in an appropriate aqueous environment.

In order to further enhance the $AuCN_2^-$ complex uptake, the addition of L-cysteine to the adsorption system was attempted [82]. Experimental results revealed that L-cysteine did enhance Au biosorption by *Bacillus*, *Penicillium*, and *Sargassum* biomass when the solution pH was lowered to pH 2. This phenomenon for pH 2 is illustrated in Fig. 12, where the ratio (R_{cys}) of gold uptake in the presence of cysteine to that without cysteine is plotted versus the final cysteine concentration. In the final cysteine concentration range (0–0.5 mM), the gold uptake by all of the three biomass types increased, following the sequence *Bacillus*, *Penicillium*, and *Sargassum* biomass. The final cysteine concentration of around 0.5 mmol/L enhanced Au uptakes by *Bacillus*, *Penicillium*, and *Sargassum* biomass up to 250% (uptake 0.02 mmol/g), 200% (0.014 mmol/g), and 148% (0.0047 mmol/g), respectively. In the presence of cysteine, the Au uptake by *Bacillus*, *Penicillium*, or *Sargassum* biomass was also found to be strongly affected by pH. The equilibrium uptakes of Au at pH 2 were greater than those at pH > 2 (Fig. 13). A similar behavior was reported for biosorption of anionic $AuCN_2^-$ by *Bacillus*, *Penicillium*, and *Sargassum* biomass without cysteine addition [81].

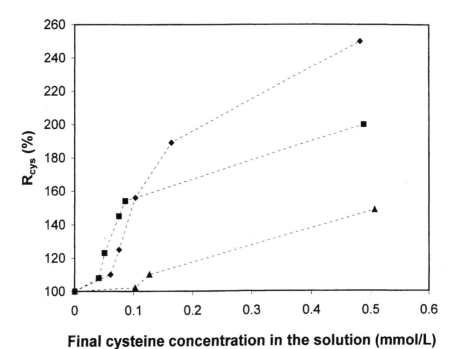

Final cysteine concentration in the solution (mmol/L)

Figure 12   Effect of L-cysteine on Au uptake by *Bacillus subtilis, Penicillium chrysogenum,* and *Sargassum fluitans* from cyanide solution. Equilibrium contact of 0.04 g biomass, initial Au concentration 0.1015 mmol/l, 20 mL solution, pH 2.0, 4 h, room temperature. ◆, *Bacillus subtilis;* ■, *Penicillium chrysogenum;* ▲, *Sargassum fluitans.* (From Ref. 82.)

     L-cysteine biosorption isotherms at pH 2 for *Bacillus, Penicillium,* and *Sargassum* biomass in Fig. 14 show encouraging uptakes by *Bacillus* and *Penicillium* biomass, while *Sargassum* biomass sorbed very little. Under the experimental conditions, the sequence for the cysteine uptake by the three biomass types was *Bacillus > Penicillium > Sargassum,* which agreed with the sequence of increased Au uptake in the presence of cysteine.

     Enhancement of Au biosorption in the presence of cysteine probably relates to the "bridging" function provided by the cysteine molecule between the gold cyanide complex and biomass. Cysteine, which figures prominently in discussions of metal ion binding to proteins, has three possible coordination sites, namely, amino, sulfhydryl, and carboxylate

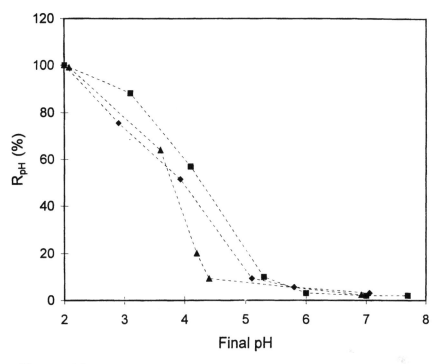

Figure 13 Effect of pH on Au uptake by *Bacillus subtilis, Penicillium chrysogenum,* and *Sargassum fluitans* from cyanide solution in the presence of cysteine. Equilibrium contact of 0.04 g biomass, 20 mL solution, initial Au concentration 0.1015 mM, initial cysteine concentration 0.51 mM, 4 h, room temperature. ◆, *Bacillus subtilis;* ■, *Penicillium chrysogenum;* ▲, *Sargassum fluitans.*

groups [95]. The dissociation constants (pKa) of those groups are, respectively, 10.36, 8.12, and 1.90 [96]. At pH 2.0, the cysteine carboxyl group is partially deprotonated and charged negatively, which enables it to combine with positively charged groups on biomass. At the same time, the cysteine amino group is protonated and charged positively, which allows for its combination with anionic $AuCN_2^-$. The peptidoglucans in *Bacillus* cell walls [97] and chitin in *Penicillium* cell walls [90] contain weak-base groups like amine or amide. As the pK (proton dissociation constant) of positively charged acetylamine groups in chitin is lower than 3.5, while amine groups of other biomolecules have pK of around 6 [98], almost

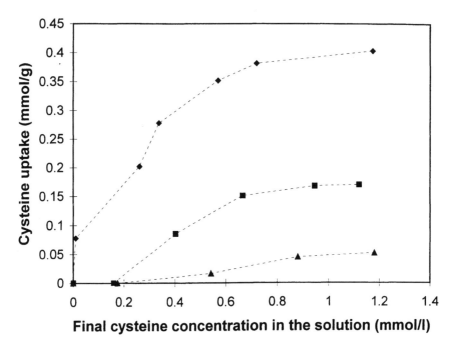

Figure 14 Cysteine biosorption isotherms by *Bacillus subtilis, Penicillium chrysogenum,* and *Sargassum fluitans.* Equilibrium contact of 0.04 g biomass, 20 mL solution, pH 2.0 4 h, room temperature. ◆, *Bacillus subtilis;* ■, *Penicillium chrysogenum;* ▲, *Sargassum fluitans.*

all amine groups on these two biomass types could be positively charged by protons at pH 2.0. This makes them amenable to combining with the carboxyl moiety on cysteine under such acidic conditions. Some carboxyl groups on cysteine may still be dissociated, as the dissociation constant (pK) of the carboxyl group on cysteine is 1.90, and the amino group is protonated and with a positive charge. This allows cysteine binding on biomass through the combination of negative cysteine carboxyl groups with some positively charged biomass functional groups. The positively charged cysteine amino groups became available for binding anionic $AuCN_2^-$. Therefore, Au could become indirectly sorbed on biomass through cysteine as a bridge: ($BFH_2^+—^-OOC—R—NH_3^+—AuCN_2^-$), where BFH_2^+ stands for the biomass functional group bearing a positive charge. The biomass, having higher affinity for cysteine than for gold cyanide, brought in extra weak-base groups for binding $AuCN_2^-$ on biomass resulting in enhanced Au uptake.

Sargassum biomass contains alginates up to 45% of its dry weight [42]. The active groups in alginates are carboxyl groups. Since the carboxyl groups of the biomass (pKa 3.5) [99] should still be protonated and therefore charged neutral at pH 2, they are less likely to contribute to either cation or anion binding. The low cysteine binding by *Sargassum* may be due rather to a smaller amount of phenolic groups known to be also present in brown seaweeds [100]. Our recent work confirmed that the presence of cysteine did increase the gold cyanide complex uptake by biomass and, in turn, the increased Au uptake was related to the cysteine-only uptake by biomass.

While cysteine presence definitely enhanced the Au biosorption uptake, the magnitude of biosorption metal anion uptake was still lower than that observed for cation biosorption [3,88,101]. This is apparently so because the sites responsible for cation binding are especially the deprotonated, negatively charged groups that often occur in larger quantities than positively charged groups. A similar behavior was observed in the anion ion exchange process, whereby a weak-base resin would have a relatively low binding capacity for anions which became indirectly attached onto active sites through proton bridges [102].

2. Ionic Strength Effect

Increasing ionic strength of the solution containing anionic sorbate species basically results in its reduced uptake, as demonstrated in studying the biosorption of Au [81,82]. Figure 15 shows the results of Au uptake as affected by the ionic strength made up by $NaNO_3$ [81,82]. As the concentration of $NaNO_3$ increased to 150 mM (ionic strength 160 mM), the uptake of Au by *Bacillus* and *Penicillium* biomass was reduced, respectively, to 64 and 46% of that without $NaNO_3$ (ionic strength 10 mM) in the solution. The Au uptake by *Sargassum* decreased almost to zero at 60 mM $NaNO_3$ (ionic strength 70 mM). Changing ionic strength (i.e., the background electrolyte concentration) influences adsorption in at least two ways: (1) by affecting the interfacial potential and therefore the activity of electrolyte ions and adsorption and (2) by affecting the competition of the electrolyte ions and adsorbing anions for available sorption sites

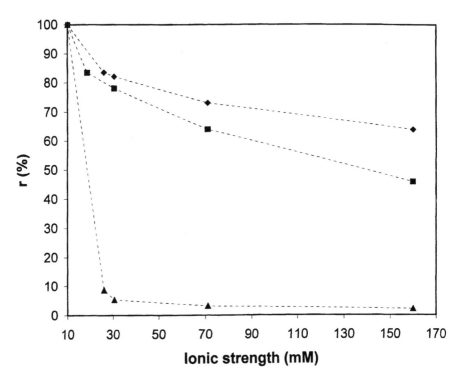

Figure 15   Effect of ionic strength on the Au relative uptake ($r = q/q_{control}$). Equilibrium contact of 0.04 g biomass, 20 mL solution, pH 2.0, initial Au concentration 20 mg/L. ◆, *Bacillus subtilis*; ■, *Penicillium chrysogenum*; ▲, *Sargassum fluitans*. (From Ref. 81.)

[86]. In our study, the added NO_3^- apparently competed with the gold cyanide complex as a counterion for the positively charged binding sites in biomass. As a result, Au uptake was reduced.

In the case with L-cysteine in the solution, increasing ionic strength also reduced the Au uptake similar to what was the case in gold cyanide adsorption by biomass only [81]. The added NO_3^- could compete with the gold cyanide complex for the positively charged binding sites on cysteine or biomass as a counterion, thus reducing the Au uptake.

3.   Desorption of Au-Loaded Biomass
The possibility of desorbing Au from biomass was examined by first sorbing Au onto biomass at pH 2 and then desorbing Au with deionized water

at pH 3, 4, or 5 adjusted by 0.1 M NaOH. Experimental results indicated that Au could be effectively eluted from Au-loaded biomass. Figure 16 shows that more than 90% of Au was recovered upon elution at pH 5 with the solid/liquid ratio [S/L (mg/mL)] = 4 for all three biomass types studied, indicating that their Au binding was easily reversible [81,103]. In accordance with the postulated weak-base mechanism of binding of the gold complex on the biomass [biomass–(H$^+$)–AuCN$_2^-$], the increasing pH shifts its sorption equilibrium allowing the Au elution.

In the case of Au uptake with the L-cysteine in the adsorption system, the AuCN$_2^-$ complex was probably bound to the cysteine protonated positively charged amino groups, while cysteine itself was bound onto biomass through the negative cysteine carboxyl group and the protonated positively charged biomass amine group. As the pH of the sorption system

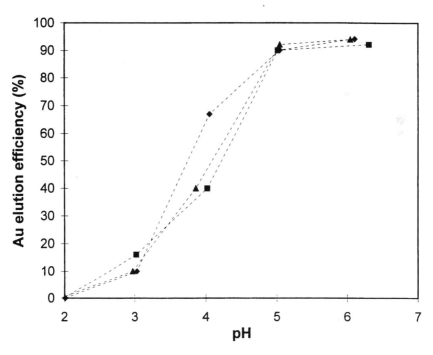

Figure 16 Effect of pH on Au elution efficiency. Equilibrium contact of 0.02 g biomass, 5 mL solution, initial Au loading was 8 μmol/g *Bacillus* biomass, 7.1 μmol/g *Penicillium* biomass, and 3.2 μmol/g *Sargassum* biomass. ◆, *Bacillus subtilis*; ■, *Penicillium chrysogenum*; ▲, *Sargassum fluitans*. (From Ref. 81.)

increased, protons dissociated from the positively charged acetyl amine groups on the biomass (pKa $<$ 3.5), thereby rendering these groups charge-neutral so that they did not attract the COO— of cysteine any longer. That led to breaking the binding "bridge" (BFH$_2^+$—$^-$OOC—R—NH$_3^+$—AuCN$_2^-$), and the gold complex became dissociated from the solid phase [82]. However, the use of concentrated NaOH leads to massive leaching of a variety of compounds from the biomass and to the destruction of its cellular structures. Therefore, Au elution was limited to the pH no higher than pH 6. These considerations point out the high stability of the gold cyanide complex, which behaved as a whole anionic complex throughout the entire sorption/desorption procedure.

4. FTIR Analysis

The FTIR analysis is a very powerful technique for establishing what groups are functional in binding the metals and in what complex form metal complex ions are likely to be. For Au biosorption by *Bacillus*, biomass FTIR confirmed that Au biosorption mainly involved complex anionic AuCN$_2^-$ species [81]. Since AuCN$_2^-$ became bound by the weak-base mechanism in an ion pair form [biomass–(H$^+$)–AuCN$_2^-$] the v(CN) gold complex peak [104] shifted only slightly to a higher position (from 2155 cm^{-1} by 15 cm^{-1}) because the bond between H$^+$ and AuCN$_2^-$ is very weak.

A weak absorbance peak at 2352 cm^{-1}, ascribed to v(NH) stretching vibrations [105] in protonated *Bacillus* biomass, disappeared on the spectrum of Au-loaded biomass, indicating that an amine group may have been involved in Au biosorption. Amine is typically a weak-base group. Primary, secondary, and tertiary amine-containing polymers are the main adsorbents used routinely for anion adsorption [102]. As AuCN$_2^-$ combined with amine in an ion pair form, the v(NH) stretching vibration was weakened because of the physical space limitation on the molecular basis, and therefore the original v(NH) biomass weak stretching vibration disappeared. The infrared analysis results confirmed that amine groups on biomass were involved in anionic gold cyanide complex adsorption through an H$^+$ bridge.

In the cysteine-aided gold biosorption by *Bacillus*, the FTIR v(NH) stretching vibrations of blank protonated *Bacillus* biomass (2313 cm^{-1}) [106] shifted for cysteine-loaded biomass (2348 cm^{-1}) [82], indicating that the amine group on biomass is most likely involved in cysteine bio-sorption. There was also a carboxyl group–related shift between the blank for pure cysteine (1595 cm^{-1}) and cysteine-loaded biomass (1651 cm^{-1}). As the pKa of COO— on cysteine was 1.9 [96], for adsorption at pH 2.0, most of the protons dissociated from carboxyl groups on cysteine, which became mainly negative —COO^{-}, were able to combine with the positively charged amine groups on the biomass. It appears that the cysteine binding may involve the combination of carboxyl groups on cysteine and amino groups on biomass. This may explain why *Sargassum* cannot effectively extract cysteine from solution at pH 2.0 as compared with *Bacillus* and *Penicillium*, since it lacks the amino groups, the main component in the *Sargassum* cell wall being alginate. The important active site of alginate is the carboxyl group, which is in its neutral form at pH 2, its pKa being 3.5 [99]. The peak of —SH was not identifiably changed (4 cm^{-1}) between the blank cysteine and cysteine-bound biomass, indicating that the —SH group on cysteine was not involved in binding cysteine to biomass.

On the spectrum of cysteine-loaded biomass and Au–cysteine-loaded biomass, the peak of (—NH) did not change, indicating that cysteine remained bound to amine groups on the biomass when Au was bound. The peak of NH$_3$$^{+}$ rocking on cysteine-loaded biomass disappeared on the spectrum of Au–cysteine-loaded biomass. These results indicate that amine groups of cysteine may be involved in Au adsorption. As the —NH$_3$$^{+}$ bound Au, possessing a high atomic weight, the rocking vibration of NH$_3$$^{+}$ disappeared.

Another peak of —SH stretching vibrations was found to shift from 2572 cm^{-1} on cysteine-only loaded biomass to 2634 cm^{-1} for Au–cysteine-loaded biomass. Sulfhydryl groups could donate a lone electron pair for the empty orbit of metal ions alone [107]. The cysteine sulfhydryl group probably co-combined AuCN$_2$$^{-}$ with the cysteine amine group through a bridging proton, whereby the amine lone electron pair becomes shared with that proton to form a relatively stable spatial structure resulting in the shifting of stretching vibrations of —SH to a higher position.

The peak of —COO^{-} asymmetric stretch on cysteine-loaded biomass did not change significantly (9 cm^{-1}) compared with that on Au–cysteine-loaded biomass, indicating that the —COO^{-} moiety was not effectively involved in Au binding. Furthermore, the spectrum of Au–cysteine–biomass featured a peak at 2280 cm^{-1}, which was ascribed to $v(CN)$ of the gold cyanide complex [104]. There was no peak of the $v(CN)$ vibration of free CN$^-$ on the spectrum of Au–cysteine-loaded biomass which should be located at 2080 cm^{-1} [104]. The FTIR results confirmed that cyanide was existing as a AuCN$_2^-$ complex in the Au-loaded biomass.

B. Biosorption of Gold Cyanide Anionic Complex Conclusions

The following conclusions can be drawn from the results and discussion presented:

1. Biosorption of metal anions and anionic complexes is preferred at lower pH.
2. The presence of "bridging" compounds such as L-cysteine can enhance anion complex biosorption at low pH (demonstrated for pH 2 with AuCN$_2^-$ and *Bacillus, Penicilium,* and *Sargassum* biomass).
3. At pH 2, the sequence of increasing anion uptake by different biomass types agreed with that of capacity for cysteine adsorption by biomass: *Bacillus* > *Penicillium* > *Sargassum.*
4. The uptake of anions may significantly decrease with increasing solution ionic strength (NaCl concentration from 0.005 M to 0.15 M).
5. Biosorption of most anions may be reversible: more than 92% of biosorbed gold cyanide complex could be eluted at pH 5.
6. The FTIR analysis confirmed that the main biomass functional groups involved in anionic gold cyanide biosorption are probably S—, N—, and O— containing groups on cysteine or on biomass.
7. Cysteine-enhanced anion (gold cyanide complex) binding apparently results from binding the anion to the cysteine NH$_3^+$ groups, while cysteine carboxyls bind to cationic groups in the biomass.

C. Biosorption of Hexavalent Chromium

Hexavalent chromium represents a particularly serious toxicity problem in effluents of some industrial operations. Its removal is notoriously difficult. In this section, the removal of hexavalent chromium by biosorption is discussed focusing on the recent work done with *Sargassum* seaweed biomass by Kratochvil et al. [78].

Very often two ionic species of Cr, namely, Cr(III) and Cr(VI), coexist in the solution. Sharma and Forster [8,76] indicated earlier the optimum Cr biosorption at pH 2 for a variety of biomasses examined, including peat moss sawdust, sugar cane bagasse, beet pulp, and maize cob. Under certain biosorption conditions hexavalent chromium is reduced to Cr(III), which behaves quite differently. Kratochvil et al. [78] noted that the binding of Cr by *Sargassum* may be responsible for the depletion of protons in the system and suggested that Cr(VI) reduction to Cr(III) was favored at lower pH value and focused on the character of interactions between hexavalent chromium and the *Sargassum* seaweed biomass. The following combined ion exchange–redox reaction mechanism for the sorption of Cr(VI) by the biomass was proposed:

$$
\text{Cr(VI) as } HCrO_4^- \left\{
\begin{array}{l}
\text{biomass is oxidized by} \quad \rightarrow \text{Cr}^{3+} \text{ and } H^+ \text{ exchange} \\
HCrO_4^- \text{ while producing Cr}^{3+} \quad \text{by cation exchange} \\
\qquad\qquad \uparrow H^+ \\
\text{biomass takes up } HCrO_4^- \\
\text{by anion exchange}
\end{array}
\right.
$$

(with Cr(VI) also as $(CrO_4^{2-}, Cr_2O_7^{2-})$)

Redox reactions:

biomass:

$$C \text{ (org.)} \rightarrow CO_2 \text{ and/or C (org.)} \rightarrow C \text{ (org. oxidized)} \tag{47}$$

chromium: $\quad HCrO_4^- + 7H^+ + 3e \rightarrow Cr^{3+} + 4H_2O \tag{48}$

Cation exchange:

$$(Cr\text{-}B_3) + 3\,H^+ \Leftrightarrow 3\,(B\text{-}H) + Cr^{3+} \tag{49}$$

Anion exchange:

$$B' + HCrO_4^- + H^+ \Leftrightarrow B'\cdot H_2CrO_4 \tag{50}$$

It should be mentioned that while all the reactions in the scheme are written for $HCrO_4^-$ ions, the mechanism can easily be extended to include the other Cr(VI) species which are likely to be present in the system, such as $Cr_2O_7^{2-}$ and/or CrO_4^{2-}. The reaction scheme suggests that $HCrO_4^-$ ions can simultaneously sorb onto biomass by ion exchange according to Reaction (50) and oxidize the biomass according to Reaction(s) (47). Furthermore, as the scheme indicates, the ion exchange and the redox reactions may proceed either sequentially and/or in parallel. In order to elucidate the role of pH in the biosorption of hexavalent chromium, the effect of pH on the equilibria of the individual Reactions (48) and (50) is discussed in the following subsections.

1. The Crucial Effect of pH

Reaction (48) describes the biomass as an anion exchanger removing Cr(VI) from aqueous solutions by the "acid adsorption" mechanism, which has previously been established as one of the mechanisms of ion exchange on weak anion exchange resins [108]. One type of weakly basic groups in the *Sargassum* biomass may be phenolic groups of polyphenolic compounds such as polyphloroglucinols which are known to be present in brown seaweed [100]. The most important feature of the acid adsorption mechanism is that in order for the uptake of chromium to occur, the liquid has to contain enough protons to effectively push the equilibrium of Reaction (50) to the right. Consequently, the equilibrium uptake of hexavalent chromium by *Sargassum* is expected to increase with decreasing pH (and vice versa). From the thermodynamic point of view, the redox Reactions (47) and (48) may proceed in the direction as they are written only if the reduction potential of $HCrO_4^-$ ions is greater than the reduction potential of the biomass, i.e., when $E(CrO_4^{2-}/Cr^{3+}) > E(B\text{-red,B-ox})$. While the value of $E(B\text{-red,B-ox})$ is generally unknown, the value of $E(HCrO_4^-/Cr^{3+})$ at 25°C can be calculated from the Nernst equation, which can be written for the Reaction (50) as follows:

$$\begin{aligned}
E(HCrO_4^-/Cr^{3+}) &= E° + \Delta E_{HCrO^-4/Cr^{3+}} + \Delta E_{pH} \\
&= E°(HCrO_4^-/Cr^{3+}) \\
&\quad + \frac{0.059}{3} \log \frac{C(HCrO_4^-)}{C(Cr^{3+})} - 7.\frac{0.059}{3} pH
\end{aligned} \tag{51}$$

Assuming isothermal conditions, Eq. (51) shows that the value of the reduction potential $E(HCrO_4^- /Cr^{3+})$ can be obtained as a sum of three terms including $E°$, $\Delta E_{HCrO_4^-/Cr^{3+}}$, and ΔE_{pH}, which reflect respective contributions of the standard reduction potential $E°(HCrO_4^-/Cr^{3+})$, the ratio of hexavalent to trivalent chromium, and the pH. The higher the ratio $C(HCrO_4)/C(Cr^{3+})$ and the lower the pH, the greater the value of the potential $E(HCrO_4^-/Cr^{3+})$, and hence the stronger the ability of $HCrO_4^-$ to oxidize the biomass.

In order to assess the respective impacts of chromium speciation and pH on the redox behavior of the system, the changes of $E(HCrO_4^-/Cr^{3+})$ caused by variations in pH, and in the ratio of $C(HCrO_4^-)/C(Cr^{3+})$, were calculated from Eq. (51) and plotted in Fig. 17. As can be seen from this figure an increase of the ratio $C(HCrO_4^-)/C(Cr^{3+})$ by two orders of magnitude produces a positive $\Delta E_{HCrO_4^-/Cr^{3+}}$ of $+39$ mV while the change of the same magnitude in pH yields a negative ΔE_{pH} of -275 mV. This means that pH has a much stronger effect on the reduction

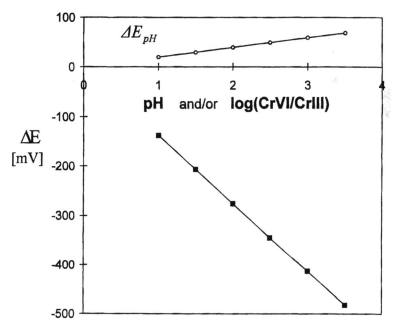

Figure 17 Effects of pH and Cr(VI)/Cr(III) ratio on the reduction potential of chromate ions calculated from the Nernst equation. $\bigcirc$, ΔE_{pH}; ■, $\Delta E_{HCrO_4^-/Cr^{3+}}$.

potential of chromate than the ratio of hexavalent to trivalent chromium. In fact, an increase of pH causes the value of $E(HCrO_4^-/Cr^{3+})$ to drop dramatically, thereby weakening the oxidizing power of $HCrO_4^-$ ions with respect to the biomass. Consequently, a critical value of pH, designated as pH_C, may be defined for any given biomass, so that at pH = pH_C, the reduction potential of $HCrO_4^-$ ions and the biomass, respectively, are equal, i.e., $E(HCrO_4^-/Cr^{3+})$ = E(B-red,B-ox). It follows from the discussion above that at pH lower than pH_C, $E(HCrO_4^-/Cr^{3+})$ > E(B-red, B-ox), and hence $HCrO_4^-$ ions oxidize the biomass, while at pH greater than pH_C, $E(HCrO_4^-/Cr^{3+})$ < E(B-red, B-ox), and all of the chromium stays in the hexavalent form.

2. Explanation of the Optimum pH Existence

At a relatively high pH, there are not enough protons to shift the equilibrium of Reaction (50) sufficiently to the right. Consequently, only a small uptake of $HCrO_4^-$ was seen at pH 4. As the equilibrium pH decreases to pH 2, the uptake of $HCrO_4^-$ by anion exchange increased. However, as soon as the equilibrium pH was lowered below the value of pH_C, $HCrO_4^-$ ions started oxidizing the biomass while producing Cr^{3+} ions. These Cr^{3+} ions then competed for the binding sites in the biomass with protons via cation exchange Reaction (49). Therefore, a further decrease of pH caused not only the value of $E(HCrO_4^-/Cr^{3+})$ to increase according to Eq. (51), thus facilitating the biomass oxidation, but also a fast desorption of Cr^{3+} from the biomass leading to lower uptakes of chromium at pH 1. Consequently, the uptake of Cr from solutions containing Cr(VI) by biomass is maximized at a value of pH which is low enough for the equilibrium of the anion exchange Reaction (50) to be shifted to the right, but still high enough to be in the vicinity of pH_C so that the reduction of $HCrO_4^-$ to Cr^{3+} does not dominate the system.

The uptake of chromium on *Sargassum* biomass at pH 7 was negligible, complemented by no significant change in the pH of the system. Furthermore, no Cr(III) was detected in the chromate-containing liquid at this equilibrium. Under acidic conditions (0.2 M H_2SO_4) almost all $HCrO_4^-$ ions present in the initial solution were reduced to Cr^{3+} over the course of 6 h, and practically no Cr(VI) was left unreacted in the systems after 1 day. Obviously, there is different kinetics at play when the Cr(VI) conversion reactions are taking place in addition to intrinsically fast ion exchange.

Kratochvil et al. [78] drew the following conclusions: At pH >2.5 Cr(III) sorbs onto *Sargassum* biomass by ion exchange in the form of $Cr(OH)^{2+}$ divalent cation. The uptake of Cr(III) by the seaweed can be modeled using Schiewer's ion exchange model [39,109]. The uptake of Cr(III) at pH <2 is negligible. The interaction between Cr(VI) and the biomass includes (1) sorption, most likely by anion exchange, and (2) reduction of Cr(VI) to Cr(III) at pH <2.5. Under highly acidic conditions (0.2 M H_2SO_4) *Sargassum* biomass is capable of effectively reducing Cr(VI) to Cr(III). The optimum pH for Cr(VI) removal by sorption lies in the region where the two mechanisms overlap, which for *Sargassum* biomass is in the vicinity of pH 2.

The results obtained under the acidic conditions have two major implications for the potential use of biosorbents: (1) it is possible to desorb most of the chromium from biomass by mineral acids via reduction to Cr(III) and (2) biosorbents may be considered as a low-cost alternative to oxidizing agents such as SO_2 and $Na_2S_2O_5$ used for the reduction of Cr(VI) to Cr(III) in the first step of the conventional treatment process. The latter suggestion is further favored by the fact that the spent plating baths are very acidic as they contain residual chromic acid.

IV. APPLICATIONS

Considering the performance of some recently identified biosorbent materials, there are extremely encouraging indications that these natural and/ or waste materials could be used particularly in situations where cheap and effective metal removal may be required. Cost-effective approach is especially important for cleaning up large volumes of dilute but toxic metal-bearing industrial effluents. Large-scale problems are notoriously associated with mining operations (acid mine drainage) where cation-bearing wastewaters abound. Metal-plating, which is a rapid-growth industry, features larges volumes of effluents containing both anions and cations. These industrial sectors represent the prime potential clients who could benefit from deployment of biosorption, emerging as a promising alternative wastewater remediation technology.

REFERENCES

1. JA Brierley. In: B Volesky, ed. Biosorption of Heavy Metals. Boca Raton, FL: CRC Press, 1990, pp 305–312.
2. I Aldor, E Fourest, B Volesky. Can J Chem Eng 73:516–522, 1995.
3. B Volesky, ZR Holan. Biotechnol Prog 11:235–250, 1995.
4. RAA Muzzarelli, F Tanfani, M Emanuelli, S Gentile. J Appl Biochem 2:380–389, 1980.
5. J Buffle, RS Altmann. In: W Stumm, ed. Aquatic Surface Chemistry. New York: Wiley, 1987, pp 351–383.
6. DW Darnall, B Greene, M Hosea, RA McPherson, M Henzl, MD Alexander. In: R Thompson, ed. Trace Metal Removal from Aqueous Solution. The Annual Chemical Congress: The Proceedings of a Symposium. London: The Royal Society of Chemistry, Industrial Division, 1986, pp 1–24.
7. N Kuyucak, B Volesky. Biotechnol Bioeng 33:823–831, 1989.
8. DC Sharma, CF Forster. Biores Technol 47:257–264, 1994.
9. ME Treen-Sears, B Volesky, RJ Neufeld. Biotechnol Bioeng 26:1323–1329, 1984.
10. RH Crist, K Oberholser, D Schwartz, J Marzoff, D Ryder, DR Crist. Environ Sci Technol 22:755–760, 1988.
11. RH Crist, K Oberholser, J McGarrity, DR Crist, JK Johnson, JM Brittsan. Environ Sci Technol 26:496–502, 1992.
12. DR Crist, RH Crist, JR Martin, J Watson. In: P Bauda, ed. Metals—Microorganisms Relationships and Applications. FEMS Symposium Abstracts, Metz, France, May 5–7. Paris: Societe Francaise de Microbiologie, 1993, 13.
13. S Schiewer, E Fourest, KH Chong, B Volesky. In: CA Jerez, T Vargas, H Toledo, JV Wiertz, eds. Biohydrometallurgical processing. Proceedings of the International Biohydrometallurgy Symposium, Santiago: University of Chile, 1995, pp 219–228.
14. N Kuyucak, B Volesky. In: B Volesky, ed. Biosorption of Heavy Metals. Boca Raton, FL: CRC Press, 1990, pp 173–198.
15. MM Figueira, B Volesky, VST Ciminelli, FA Roddick. Wat Res 34:196–204, 2000.
16. B Volesky, ed. Biosorption of Heavy Metals. Boca Raton. FL: CRC Press, 1990.
17. E Guibal, C Roulph, P Le Cloirec. Wat Res 26:1139–1145, 1992.
18. NA Yakubu, AWL Dudeney. In: HH Eccles, S Hunt, eds. Immobilisation of Ions by Biosorption. Chichester, England: Ellis Horwood, 1986, pp 183.

19. M Galun, P Keller, H Feldstein, S Siegel, B Siegel. Water Air Soil Pollut 20:221, 1983.

20. R Jilek, H Prochazka, I Kuhr, J Fuska, P Nemec, J Katzer. Czechoslovakian patent 155, 830, 1971.

21. MZ-C Hu, JM Norman, BD Faison, M Reeves. Biotechnol Bioeng 51: 237–247, 1996.

22. NDH Munroe, JD Bonner, R Williams, KF Pattison, JM Norman, BD Faison. Abstracts, 15th Symposium on Biotechnology for Fuels and Chemicals, May 10–14. Oak Ridge, TN:ORNL, 1993.

23. LE Macaskie, RM Empson, AK Cheetham, CP Grey, AJ Skarnulis. Science 257:782–784, 1992.

24. N Friis, P Myers-Keith. Biotechnol Bioeng 28: 21–28, 1986.

25. GW Strandberg, SE Shumate II, JR Parrot. Appl Environ Microbiol 41: 237–245, 1981.

26. H May. Biosorption by Industrial Microbial Biomass. M Eng thesis. McGill University, Montreal, Canada, 1984.

27. B Volesky, HA May-Phillips. Appl Microbiol Biotechnol 42:797–806, 1995.

28. JJ Byerley, JM Scharer, AM Charles. Chem Eng J 36:B49–B59, 1987.

29. T Horikoshi, A Nakajima, T Sakaguchi. Agric Biol Chem 332:617, 1979.

30. DN Edgington, SA Gorden, MM Thommes, LR Almodovar. Limnol Ocean 15:945–955, 1970.

31. B Volesky, I Prasetyo. Biotechnol Bioeng 43:1010–1015, 1994.

32. M Tsezos, B Volesky. Biotechnol Bioeng 23:583–604, 1981.

33. ZR Holan, B Volesky. Biotechnol Bioeng 43:1001–1009, 1994.

34. S Schiewer, B Volesky. Environ Sci Technol 30:2921–2927, 1996.

35. J Yang, B Volesky. Environ Sci Technol (submitted, 2000).

36. CFJ Baes, RE Mesmer. The Hydrolysis of Cations. New York: Wiley-Interscience, 1976, pp 197–240.

37. W Mackie, RD Preston. In: WDP Stewart, ed. Algal Physiology and Biochemistry. Oxford, UK: Blackwell Scientific Publications, 1974, pp 58–64.

38. J Buffle. Complexation Reactions in Aquatic Systems: An Analytical Approach. Chichester, England: Ellis Horwood, 1988, pp 163–173.

39. S Schiewer, B Volesky. Environ Sci Technol 29:3049–3058, 1995.

40. CGJ Hill. An Introduction to Chemical Engineering Kinetics and Reactor Design. New York: Wiley, 1977, pp 167–204.

41. WD Schecher. MINEQL+: A Chemical Equilibrium Program for Personal Computers, Users Manual Version 2.22. Hallowell, ME: Environmental Research Software, Inc., 1991.

42. E Fourest, A Serre, J-C Roux. Toxicol Environ Chem 54:1–10, 1996.
43. J Ferguson, B Bubela. Chem Geol 13:163–186, 1974.
44. JM Tobin, DG Cooper, RJ Neufeld. Appl Environ Microbiol 47:821–824, 1984.
45. DW Darnall, B Greene, MT Henzl, JM Hosea, RA McPherson, J Sneddon, MD Alexander. Environ Sci Technol 20:206–208, 1986.
46. B Greene, R McPherson, D Darnall. In: JW Patterson, R Pasino, ed. Metals Speciation, Separation and Recovery. Chelsea, MI: Lewis, 1987, pp 315–338.
47. GJ Ramelow, D Fralick, Y Zhao. Microbiol 72:81–93, 1992.
48. ZR Holan, B Volesky, I Prasetyo. Biotechnol Bioeng 41:819–825, 1993.
49. B Greene, M Hosea, R McPherson, M Henzl, MD Alexander, DW Darnall. Environ Sci Technol 20:627–632, 1986.
50. RH Crist, JR Martin, J Chonko, DR Crist. Environ Sci Technol 30:2456–2461, 1996.
51. P Brassard, E Macedo, S Fish. Environ Sci Technol 30:3216–3222, 1996.
52. D Dobos. A Handbook for Electrochemists in Industry and Universities. Amsterdam: Elsevier Scientific, 1994, p 88.
53. AL Horvath. Handbook of Aqueous Electrolyte Solutions. Chichester, England: Ellis Horwood, 1985, p 289.
54. RH Crist, JR Martin, D Carr, JR Watson, HJ Clarke, DR Crist. Environ Sci Technol 28:1859–1866, 1994.
55. L Lapidus, GE Pinder. Numerical Solution of Partial Differential Equations in Science and Engineering. New York: Wiley, 1982.
56. F Helfferich. Ion Exchange. New York: McGraw-Hill, 1962, pp 72–94.
57. B Westrin, A Axelsson. Biotechnol Bioeng 38:439–446, 1991.
58. E Fourest, B Volesky. Appl Biochem Biotechnol 67:33–44, 1997.
59. JD Dodge. The Fine Structure of Algal Cells. London: Academic Press, 1973, pp 14–45.
60. E Percival, RH McDowell. Chemistry and Enzymology of Marine Algal Polysaccharides. London: Academic Press, 1967, pp 127–156.
61. JB Rosen. J Chem Phys 20:387–394, 1952.
62. ML Apel, AE Torma. In: AE Torma, ML Apel, CL Brierley, eds. Biohydrometallurgical technologies. Proceedings of the International Biohydrometallurgy Symposium, Warrendale, PA: The Minerals, Metals & Materials Society, 1993, pp 25–34.
63. M Tsezos, SH Noh, MHI Baird. Biotechnol Bioeng 32:545–553, 1988.
64. JC Crittenden, WJJ Weber. J Environ Eng Div ASCE 104:1175–1195, 1978.

65. JC Crittenden, AM Asce, WJ Weber, Jr. J Environ Eng Div 104:185–197, 1978.
66. JC Crittenden, AM Asce, WJ Weber, Jr. J Environ Eng Div 104:433–443, 1978.
67. KT Liu, WJ Weber. J Water Poll Control Fed 53:1541–1550, 1981.
68. J Villadsen, M Michelsen. Solution of Differential Equation Models by Polynomial Approximation. Englewood Cliffs, NJ: Prentice-Hall, 1978, pp 111–143.
69. D Kratochvil, B Volesky, G Demopoulos. Wat Res 31:2327–2339, 1997.
70. H Karberry. A I Ch E J 6:460–462, 1960.
71. JE Williamson, KE Bazaire, CJ Geankoplis. Ind Eng Chem Fundam 2:126–129, 1963.
72. T Kataoka, H Yoshida, K Ueyama. J Chem Eng Japan 5:132–136, 1972.
73. T Furusawa, JM Smith. Ind Eng Chem Fundam 12:197–203, 1973.
74. WJ Weber, KT Liu. Chem Eng Commun 6:49–60, 1980.
75. C Milot, E Guibal, J Roussy, P LeCloirec. In: FM Doyle, N Arbiter, N Kuyucak, eds. Mineral Processing and Extractive Metallurgy Review. Newark, NJ: Gordon and Breach, 1997, pp 293–308.
76. DC Sharma, CF Forster. Wat Res 27:1201–1208, 1993.
77. S Bosinco, J Roussy, E Guibal, P LeCloirec. Environ Technol 17:55–62, 1996.
78. D Kratochvil, P Pimentel, B Volesky. Environ Sci Technol 32:2693–2698, 1998.
79. B Volesky, N Kuyucak, US patent no. 4,769,233, 1988.
80. N Kuyucak, B Volesky. Biorecovery 1:189–204, 1989.
81. H Niu, B Volesky. J Chem Technol Biotechnol 74:778–784, 1999.
82. H Niu, B Volesky. J Chem Technol Biotechnol 75:436–442, 2000.
83. J Marsden, L House. Chemistry of Gold Extraction. Hartnoll, England: Ellis Horwood, 1993, pp 100–500.
84. AD Eaton, LS Clesceri, AE Greenberg, eds. In: Standard Methods for the Examination of Water and Wastewater. Washington: American Public Health Association, 1995, Ch 4: pp 19–71.
85. NCM Gomes, VR Linardi. Revista Microbiol (Port) 27:218–222, 1996.
86. C Tien. Adsorption Calculations and Modeling. Boston: Butterworth-Heinemann, 1994, pp 29–40.
87. J Remacle. In: B Volesky, ed. Biosorption of Heavy Metals. Boca Raton, FL: CRC Press, 1990, pp 83–92.
88. BJ Fen, CJ Daughney, N Yee, TA Davis. Geochim Cosmochim Acta 61:3319–3328, 1997.

89. J Buffle. Complexation Reactions in Aquatic Systems: An Analytical Approach. Chichester, England: Ellis Horwood, 1988, pp 195–303.
90. FA Troy, H Koffler. J Biol Chem 244:5563–5576, 1969.
91. E Fourest, B Volesky. Environ Sci Technol 30:277–282, 1996.
92. KF Hayes. J Coll Interf Sci 115:564–572, 1987.
93. VP Harden, JO Harris. J Bacteriol 65:198–202, 1953.
94. W Stumm. Chemistry of the Solid–Water Interface. New York: Wiley, 1992, pp 43–81.
95. H Shindo, TL Brown. J Amer Chem Soc 87:1904–1908, 1965.
96. DD Perrin. Stability Constants of Metal-Ion Complexes. Oxford, England: Pergamon Press, 1979.
97. TJ Beveridge, RGE Murray. J Bacteriol 127:1502, 1976.
98. GAF Roberts. Chitin Chemistry. London: Macmillan, 1992, pp 204–206.
99. S Schiewer, B Volesky. Environ Sci Technol 31:2478–2485, 1997.
100. M Ragan, JS Craigie. In: JA Hellebust, JS Craigie, eds. Handbook of Phycological Methods. New York: Cambridge University Press, 1978, pp 157–179.
101. H Niu, XS Xu, JH Wang, B Volesky. Biotechnol Bioeng 42:785–787, 1993.
102. JA Marinsky. Ion Exchange. New York: Marcel Dekker, 1966, pp 228–231.
103. H Niu, B Volesky, N Gomes. In: R Amils, A Ballester, eds. Biohydrometallurgy and the Environment Toward the Mining of the 21st Century (Part B): International Biohydrometallurgy Symposium—Proceedings. Amsterdam, The Netherlands: Elsevier Science, 1999, pp 493–502.
104. K Nakamoto. Infrared and Raman Spectra of Inorganic and Coordination Compounds. New York: Wiley, 1986, p 231.
105. MSC Flett. Characteristic Frequencies of Chemical Groups in the Infra-Red. New York: Elsevier, 1963, p 24.
106. EGJ Brame, JG Grasselli. Infrared and Raman Spectroscopy. New York: Marcel Dekker, 1977, p 745.
107. NC Li, RA Manning. J Amer Chem Soc 77:5225–5528, 1955.
108. R Kunin. Ion Exchange Resins. New York: Wiley, 1958, pp 73–105.
109. S Schiewer. J Appl Phycol 11:79–87, 1999.
110. J Crank. Mathematics of Diffusion, 2nd ed. London, UK: Clarendon Press, 1975, p 94.

5

Synthesis and Application of Functionalized Organo-Ceramic Selective Adsorbents

Lawrence L. Tavlarides and J. S. Lee

Syracuse University, Syracuse, New York

I. INTRODUCTION

Traditionally, solvent extraction involves contacting of an aqueous and organic phase in order to extract solutes such as metal ions or biochemical products to separate them from mixtures or to concentrate them. The organic "solvent" may contain a chelation ligand or ion exchanger in a diluent. Along with the contactor, process equipment requires fluid phase separators to separate the loaded organic extractant from the aqueous feed. Then a stripping step is required to recover the metal ion/biochemical and, perhaps, regeneration of the solvent. Such liquid extraction processes have wide applicability for high-volume hydrometallurgical, radionuclear, and biochemical separations.

An alternative approach is to immobilize the chelating or ion exchange ligands to a polymeric or inorganic solid support. The former supports consist of solid polymer networks within which are embedded the exchange ligands. Another form is an aqueous phase soluble polymer to which the ligand is attached. The class of inorganic support we will discuss are silica-based materials. These can be classified into two categories: (1) preformed silica gels onto which ligands are placed or (2) organo-

169

ceramic polymers with the ligand embedded within the ceramic matrix. We will limit this discussion to metal ion separations for which these materials find application in hydrometallurgical processing for finishing steps, electrochemical metal refining, electroplating operations, and remediation of acid mine drainage solutions. The topics which will be discussed here are selection of ligands, synthesis of extractants immobilized on inorganic materials, characterization, adsorption isotherms, kinetics of adsorption, mass transfer characteristics, and applications.

II. SYNTHESIS OF SOLID SUPPORTED EXTRACTION MATERIALS

A. Classification of Solid Supported Extractants

The solid supported extractants can be classified based on the support matrix, such as organic polymers and inorganic oxides. To choose a proper support matrix, one should consider the availability of the synthesis chemistry, the simplicity of the synthesis, the stability of the structure, the allowance of proper pore structure for rapid mass transfer in the support matrix, and economic aspects.

The extractants can be categorized also according to whether they are ion exchangers or metal chelating agents, as shown in Fig. 1 [1]. The capability of metal ion extraction by a solid extractant mainly depends on the nature of the functional groups such as acid–base characteristics, polarizability, and molecular structure of the functional ligand, even though the solid matrix geometry is also important.

For selective metal ion separations, coordinating chelates are favorable. The chelating extractants can be further categorized based on the functional atoms in the functional ligands. Pearson [2] proposed hard–soft acid–base theory in which the donor atoms such as oxygen, nitrogen, and sulfur are classified as hard, intermediate, and soft bases, and metal ions as hard, intermediate, and soft acids. The classification of metal ions is shown in Table 1.

According to the theory, the coordination easily occurs between hard acid and hard base, and soft acid and soft base. Hard metal ions such as Fe^{3+}, Co^{3+}, and UO_2^{2+} easily form coordination with the molecules that have oxygen atoms. On the other hand, soft metal ions such as Pt^{2+}, Cd^{2+},

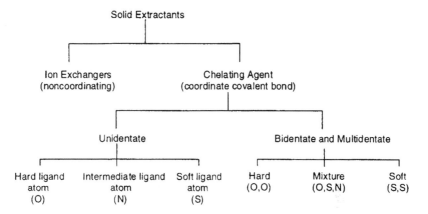

Figure 1   The classification of solid extractants based on functional ligand and/or donor atoms. (From Ref. 1.)

and Hg^{2+} have a tendency of complex formation with sulfur-containing ligands instead of with oxygen-containing ligands. For example, Khisamutdinov et al. [3] showed the complex formation of 2,5-bis-(butylthiomethyl)-ethylenethiourea with platinum(II), as shown in Fig. 2.

The functional ligands in solid extractants frequently contain polyfunctional groups instead of one functional atom. The multiple functionalities in a functional ligand can provide environments in which very specific metal–ligand complexes are formed. For example, Chessa et al. [4] showed the extraction of palladium chloride from a mixture of palladium, platinum, and gold chloride in aqueous solutions with a dipicolinic acid

Table 1   Classification of Lewis Acids

Hard	Borderline	Soft
H^+, Li^+, Na^+, K^+, Be^{2+}, Mg^{2+}, Ca^{2+}, Sr^{2+}, Sn^{2+}, Al^{3+}, Sc^{3+}, Ga^{3+}, In^{3+}, La^{3+}, Cr^{3+}, Co^{3+}, Fe^{3+}, As^{3+}, Ir^{3+}, UO_2^{2+}	Fe^{2+}, Co^{2+}, Ni^{2+}, Cu^{2+}, Zn^{2+}, Pb^{2+}	Cu^+, Ag^+, Au^+, Tl^+, Hg^+, Cs^+, Pd^{2+}, Cd^{2+}, Pt^{2+}, Hg^{2+}, CH_3Hg^+, Ti^{3+}, I^+, Br^+, M_o (metal atoms)

Source: Ref. 2. Courtesy of the American Chemical Society.

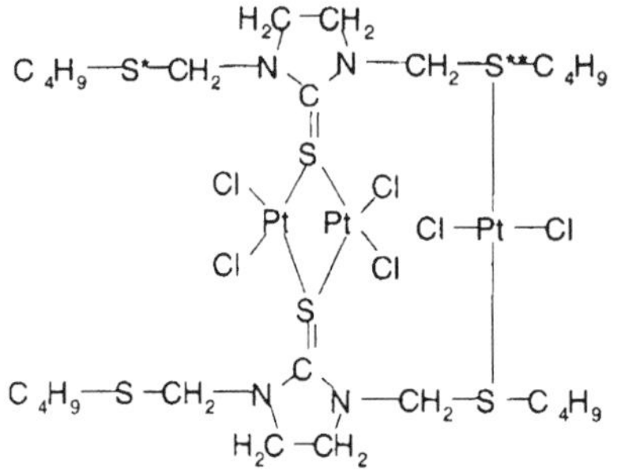

Figure 2 Complexes of platinum(II) chloride with thiourea moiety. Only one of the sulfur atoms of the sulfide group (S** or S*, with equal probability) takes part in the coordination with the platinum atom, which can be explained by steric hindrances. (From Ref. 3.)

immobilized resin. The dipicolinic acid molecule contains oxygen and nitrogen atoms and is very selective for palladium chloride over platinum and gold chloride in the pH range of 0 to 7.0, as shown in Fig. 3.

B. Synthesis of Solid Extractants

1. Polymeric Extractants

Major efforts have been directed to the synthesis of organic polymer resins and in process developments using these resins for metal ion extractions over the last 50 years. Hence, a variety of polymeric extractants, either ion exchangers or chelating resins, are available and the synthesis techniques are well established [1,5,6]. Polymeric metal extractants can be easily tailored and adopted for specific applications, such as hydrometal-

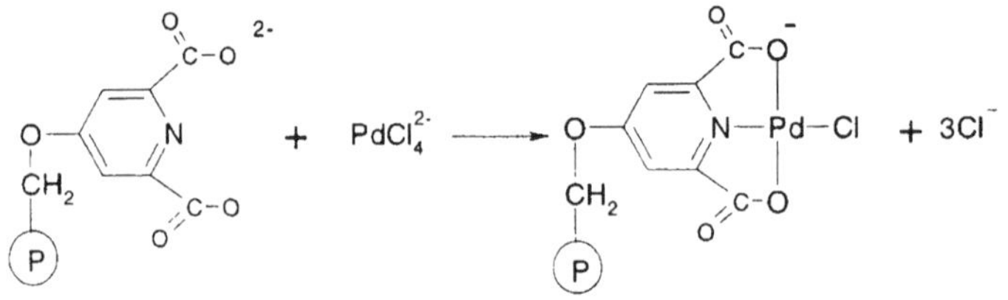

Figure 3 A complex formation of $PdCl_4^{2-}$ with a dipicolinic acid polystyrene-based chelating resin. (From Ref. 4.)

lurgical metal separations, nuclear waste treatments, and electroplating wastes cleanup. Recently, water-soluble polymers have attracted attention as one of the alternatives of polymeric solid resins. The water-soluble polymer system consists of metal extraction with water-soluble polymers, the separation of the polymers from the aqueous solution by ultrafiltration, and regeneration of the polymers. An attractive feature of this system includes the elimination of the diffusive mass transfer resistance which exists in solid supports. Hence, the extraction rate is much faster than the solid resin systems [7,8].

The details of the polymeric extractant systems are beyond the scope of this discussion and will not be presented here.

2. Organo-Ceramic Extractants

Organo-ceramic extractants made of functional ligands and inorganic supports are attractive because of the mechanical strength, thermal stability, nonswelling structure, wide range of particle size, and well-defined pore structure of these materials. The well-defined pore structure in particular provides a good environment for diffusion of metal ions in the material matrix. Also, it is reported that these types of materials are chemically stable over a wide range of acidity, i.e., [HCl] > 6.0 mol/L up to pH 11 or 12 [9]. Especially it is resistant to alkaline hydrolysis even though the skeleton of the adsorbent is made of hydrous oxide. This resistance is largely presumed a result of the nonwettable property of the silica network protected by organic chains as parts of the attached ligands.

The organically modified hydrous oxides, especially in adsorptive separation fields, have been applied as packing materials for high-performance liquid chromatography in the 1970s and 1980s [10–13]. These materials have been reported in the literature in recent years [14]. Earlier attempts to synthesize surface-modified oxide materials were not successful as adsorbents for metal separation processes [9]. Many examples of the surface-modified silica and its applications for metal extractions in the 1980s were reported in the literature and well summarized [15]. Continued attempts to synthesize surface-modified oxide adsorbents have been made in the 1990s and showed promising results for many applications [9,16–22]. Recently, it was reported that the use of meso-porous silica as a support for the immobilization of a propylthio moiety could improve metal uptake capacity [23,24].

In most cases of the organo-ceramic adsorbents, silica oxide has been used as a support material because of the ease of the texture modification [32] and its mechanical and thermal stabilities [46].

A variety of methods have been employed to synthesize these adsorbents. These methods can be classified into three basic categories, even though the synthetic routes within a category may vary: (1) solvent deposition, (2) covalent attachment of the ligand, and (3) formation of organo-ceramic polymers. The former two methods are considered as conventional methods for the synthesis of organo-ceramic adsorbents to this day. The latter is a recently emerging synthesis technique.

a. Solvent Deposition Method A variety of solvent extractants have been developed and used in extractive metal separation processes. These solvent extractants have normally high selectivity for specific metal ions depending on the condition of the aqueous solution matrix, such as solution pH and types of counterions. However, the solvent extraction method is not favorable economically for the treatment of dilute solutions [25]. Hence, the impregnations of such extractants on polymer resins [26,27] and inorganic supports [28–30] have been proposed as an alternative.

The general approach of solvent deposition is composed of two sequential steps, surface modification with a silane coupling agent and solvent deposition on the modified silica. A schematic of solvent-deposited adsorbent is shown in Fig. 4. The surface modifications of hydrous oxide with coupling agents are possible through coupling reactions between the hydroxyl groups of the support oxides and coupling agents such as dimethyl-dichloro-silane,

$$2\equiv Si-OH + Cl_2Si(CH_3)_2 \rightarrow (Si-O)_2-Si-(CH_3)_2 \ldots$$

Polar or nonpolar organo-silane coupling agents are available commercially [31]. Details of silane coupling agents such as dimethyl-dichloro-silane, octyltriethoxysilane, and phenyltriethoxysilane used for surface modification of silica are discussed elsewhere [11]. A suitable chain length for the functional groups can be selected by considering pore size and pore volume of the oxides.

When the solvent extractant is deposited on the modified silica by evaporation from a solution, the carbon chain of the extractant is retained on the bonded organic chains through van der Waals forces. In order to prepare the solvent-deposited adsorbents with high stability and capacity

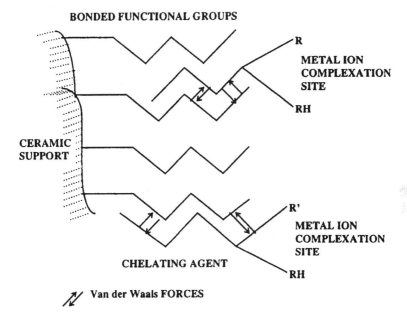

Figure 4 Inorganic chemically active adsorbent prepared by solvent deposition of chelating agents on the functionalized surface. (From Ref. 28.)

and kinetic rates equal or greater than pore diffusion rates, the properties that the chelating agents should possess include [28]

1. Very low solubility in water
2. Hydrocarbon chains away from the complexing moiety to retain hydrophilicity at the complexing end and to prevent steric hindrance to the formation of chelate rings
3. Sufficient thermal stability so that extractant is not destroyed or altered during immobilization since the mixture is heated to remove excess solvent
4. Sterically compact geometry comparable with the pore size and pore volume of functionalized support so that the extractant can penetrate into the pores and interact with the bonded functional groups
5. Sufficient chemical stability so that it retains its activity during operation

Some adsorbents synthesized through the solvent deposition method are summarized and shown in Table 2.

Table 2 Surface-Modified and Solvent-Deposited Inorganic Adsorbents

Support	Cheating agent	Metal ions	Applications/comments
Silica gel alumina	Kelex-100[®a]	Pb(II), Cu(II), Ni(II), Cd(II)	Removal Pb(II) to less than 1 ppm and stable for 20 adsorption/stripping cycles
Silica gel	Lix-84[®b]	Cu(II), Ni(II)	Removal of Cu(II) and Ni(II) to less than 1 ppm from acidic streams in electroplating industries
Silica gel	Cyanex-272[®c]	Cd(II), Zn(II), Co(II), Ni(II)	Removal of Cd(II) to less than 2 ppm
Silica gel	Cyanex-302[®c]	Cd(II), Zn(II), Pb(II)	Simultaneous removal and separation by selective stripping
Silica gel	DEHPA[®d]	Cd(II), Pb(II), Nd(III), and rare earths	Removal and preconcentration of Nd(III) from dilute (10 ppm) solutions
Silica gel	Alamine[b]	Cr(VI) and Cu(CN)$_2$, Cu(N)$_3^{2-}$	Recovery and recycle of copper cyanide from cyanide waste solutions

[a] Sherex Chemical Co.
[b] Henkel Co.
[c] American Cyanamide Co.
[d] Di-2-ethylexyl-phosphoric acid—Johnson Mattey.
Source. Ref. 28.

b.   Covalent Attachment to the Support   The covalent attachment technique for the functionalization of inorganic supports is a very elegant approach in the synthesis of adsorbents. This synthesis method can produce organo-ceramic adsorbents with higher stability, selectivity, and adsorption rates.

The functional precursors used in this synthesis have bifunctionality. One end of the precursor ligand is either a chlorosilane or an alkoxysilane, which reacts with hydroxyl groups on the oxide surface. The other end is a functional group active for metal ions. These two ends are bridged mostly with alkyl chains. The $\equiv$Si—C bond is fairly stable from hydrolysis reactions in aqueous solutions [11]. The $\equiv$Si—O—Si$\equiv$ bond that composes the skeleton of the silica is easily dissolved at high pH (pH > 10). Hence, uniform coverage of the surface of silica with the functional precursor molecule is important for the adsorbent stability. Also, it was proposed that titanium oxide coatings prior to surface modification with the functional precursors could prevent dissolution of the surface in highly caustic solutions [21].

The functional groups taking part in the formation of metal chelates usually include nitrogen, oxygen, phosphorus, and sulfur atoms. These donor atoms are capable of reacting with metal ions, owing to the coordinate/covalent/ionic bond. Metal chelates can be formed with the participation of donor atoms situated in one unit of matrix or at the matrix chain. Accordingly, highly selective surface-modified ceramic adsorbents can be prepared by careful planning and execution of synthesis schemes to introduce desired donor atoms in a preferred geometry.

A schematic structure of the adsorbents synthesized through covalent attachment is shown in Fig. 5. Numerous studies have been performed for the synthesis and applications of these types of adsorbents. Some of the adsorbents found in the literature are given in Table 3 as examples.

The synthesis methods used for covalent attachment of ligands can be divided into two basic methods even though the details of each approach may vary.

In the first approach a coupling agent is first attached to the support surface and then a functional organic molecule is attached to the lattice, as outlined in Fig. 6. Also, the functional ligands immobilized on inorganic supports may be further modified to provide better metal selectivity for target metal ions or to modify the acid–base properties of the ligands.

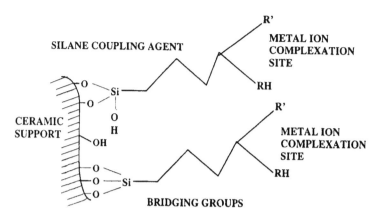

Figure 5 Inorganic chemically active adsorbent prepared by covalent bonding using silane-coupling agent. (From Ref. 28.)

For example, Soliman [46] synthesized modified silica sorbents with a series of aliphatic amines and conducted further modification of the amines with isatin (2,3-indolinedione) and ninhydrin (1,2,3-indantrione monohydrate) reagents. The author observed the addition of isatin and ninhydrin reagents apparently improved the uptake capacity of Fe(III) compared to the amine immobilized sorbents.

In the second approach, a functionalized silane coupling agent is synthesized by reacting the silane-coupling agent with functional groups. Then, the functionalized silane coupling agent is attached to the silica surface. This procedure is outlined in Fig. 7.

In general, it was observed that, method B produces high-capacity materials due to high coverage density of functional groups and homogeneous attachment. However, in some instances, functionalization of the silane coupling agent is not possible in the liquid phase.

A recent advancement in the covalent attachment is the use of meso-porous silica as the support. Meso-porous silicas have high surface area and reasonable pore diameter and pore volume. It was reported that the use of hexagonal meso-porous silica materials [23,24,49] with high surface area (up to 1500 m²/g) and well-defined uniform pore size could produce a high-performance mercury adsorbent. In the synthesis of the adsorbent, multiple silanizations were performed to increase the ligand density on

Table 3 ICAAs Prepared by Covalent Bonding

Support	Functional group	Method/coupling agent	Metal ions	Ref.
Silica	8-Hydroxyquinoline	Method A/γ-aminopropyltriethoxysilane	Cd(II), Zn(II), Co(II), Ni(II)	12
Silica	NH$_2$, NHCH$_2$CH$_2$NH$_2$, NH(CH$_2$CH$_2$NH)$_4$H, CH(COCH$_3$)$_2$, N(CH$_2$COOH)$_2$	Not reported/not reported	Cu(II), Co(II), Zn(II), Fe(III), Ni(II), Mn(II)	32
Silica	8-Hydroxyquinoline	Method A/aminophenyltrimethoxysilane	Cu(II)	13
Silica	Allyloxymethyl-18-crown-6, 15-crown-5	Method B/diethoxymethylsilane	Sr(II), Ba(II), Cd(II), Tl(I), Ag(I)	33
Silica	Hydroxamic Acid Groups	Method A/3-trichlorosilylpropionate	Mo(VI), Zr(IV), V(V)	34
Silica	1,5-Dithia-19-crown-6, 1,4,7,10-tetrathia-18C6	Same as Ref. 33	Pd(II), Au(III), Ag(I), Hg(II)	16
Silica	Thiol, amines, phosporus, and its derivatives	Not reported/not reported	Ag(I), As(III), Sb(III), Bi(III), Cu(II), Ni(II), Ga(III), Fe(III), Pd(II), Pt(IV)	9

Table 3 Continued

Support	Functional group	Method/coupling agent	Metal ions	Ref.
Silica	Iminodiacetic acid	Not reported/not reported	Fe(II), Cu(II), Pb(II), Co(II), Ni(II), Zn(II), Cd(II), Mn(II)	35
Silica	Imidazole derivatives	Method A/(3-glycidoxypropyl)trimethoxysilane	Cu(II), Ni(II), Co(II), Cd(II), Zn(II)	36
Silica	Dithiocarbamate	Method A/3-chloropropyltrimethoxysilane	Co(II), Ni(II), Cu(II), Zn(II) in EtOH solutions	37
Silica	N-(2-pyridyl)acetamide	Method A/3-chloropropyltrimethoxysilane	Ni(II), Co(II), Cu(II), Zn(II), Cd(II), Hg(II) in ethanol and propanone	38
Silica	Diphosphonic acid	Method A/trichlorovinylsilane	Actinide elements	39
Silica	5-Methyl-8-hydroxyquinoline	Method A & B/aminopropyltriethoxysilane	Pb(II), Cu(II), Ni(II), Cd(II)	28
Silica	Thio, sulphide acid	Method A/3-mercaptopropyltrimethoxysilane	Cd(II), Pb(II), Zn(II)	28
Silica	Primary, secondary, and tertiary amines, and diazole	Method A/aminopropyltryethoxysilane or 3-chloropropyltrimethoxysilane	Cu(II), Ni(II), anionic complexes, $Cr_2O_7^{2-}$, CrO_4^{2-}	28
Silica	Phospho acids and aminophospho acids	Method A/aminophenyltrimethoxysilane	Antimony(V), Cd(II), Zn(II), Nd(III), and rare earths	28

Support	Ligand	Method	Metal ions	Ref.
Silica	Pyrogallol	Method A/aminophenyltrimethoxysilane	Sb(III), Al(III), Cu(II)	28
Silica	Bis(2-pyridylakyl)amines	Method B/(3-glycidoxypropyl)trimethoxysilane	Cu(II), Ni(II), Co(II), Cd(II), Zn(II)	40
Silica	2,5-Dimercapto-1,3,4-thiadiazole	Method A/3-chloropropyltrimethoxysilane	Zn(II), Cd(II), Ni(II), Pb(II), Co(II), Fe(III)	41
Silica	Thiol	Method A/3-mercaptopropyltrimethoxysilane	Hg(II)	42
Silica	Thiol	Method A/3-mercaptopropyltrimethoxysilane	Hg(II)	23
Silica	5-Amino-1,3,4-thiadiazole-2-thiol	Method A/3-chloropropyltrimethoxysilane	Cu(II) in acetone and EtOH solutions	43
Silica	3-Amino-1,2,4-triazole	Method A/3-chloropropyltrimethoxysilane	Cu(II), Co(II) in EtOH and acetone solutions	44
Silica	2-Amino-1,3,4-thiadiazole	Method A/3-chloropropyltrimethoxysilane	Cu(II) in acetone and EtOH solutions	45
Silica	Thiol	Method A/3-mercaptopropyltrimethoxysilane	Hg(II)	24
Silica	Alkyl amines and its derivatives with isatin and ninhydrin	Method A/3-aminopropyltrimethoxysilane or 3-chloropropyltrimethoxysilane	Fe(II), Co(II), Ni(II), Cu(II), Pb(II), Cd(II), Zn(II)	46
Silica	Thiazolidine-2-thione	Method A/3-chloropropyltrimethoxysilane	Cu(II) in acetone and EtOH solutions	47
Silica	2-Aminothiazole	Method A/3-chloropropyltrimethoxysilane	Cu(II) EtOH and acetone solutions	48

Step 1: Immobilization of Coupling Agent

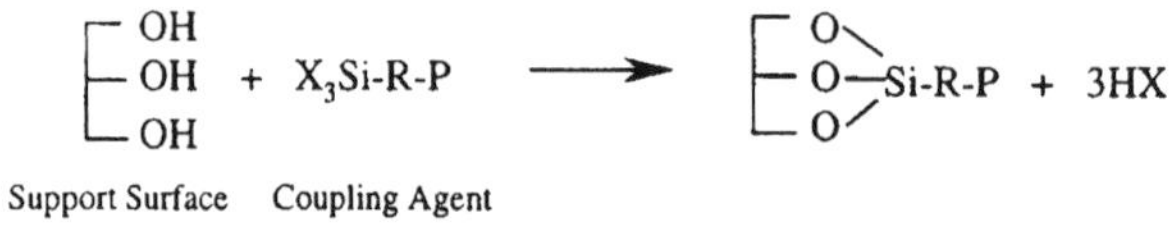

$$
\begin{bmatrix} \text{OH} \\ \text{OH} \\ \text{OH} \end{bmatrix} + X_3\text{Si-R-P} \longrightarrow \begin{bmatrix} \text{O} \\ \text{O} \\ \text{O} \end{bmatrix}\!\!\text{Si-R-P} + 3HX
$$

Support Surface   Coupling Agent

> X: Halide, Alkoxy, Acetoxy, and/or Hydroxy
> R: Substituted or Unsubstituted Alkyl/Aryl
> P: Appropriate Reactive Group

Step 2: Ligand Attachment

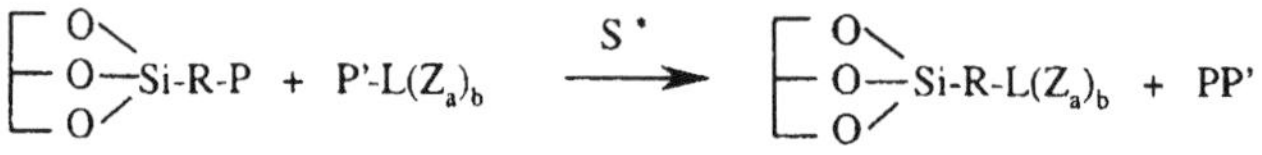

$$
\begin{bmatrix} \text{O} \\ \text{O} \\ \text{O} \end{bmatrix}\!\!\text{Si-R-P} + \text{P'-L}(Z_a)_b \xrightarrow{\ S^{\,*}\ } \begin{bmatrix} \text{O} \\ \text{O} \\ \text{O} \end{bmatrix}\!\!\text{Si-R-L}(Z_a)_b + PP'
$$

> P': Appropriate Reactive Group
> Za: Donor Atom of Type 'a'
>   a = 1 - 8 (upto eight) types
> b: Number of each Donor Atom per Ligand
> *: Different Reaction Schemes to Attach Ligand

Figure 6   Covalent attachment to support by method A.

Step 1: Ligand Attachment to Coupling Agent

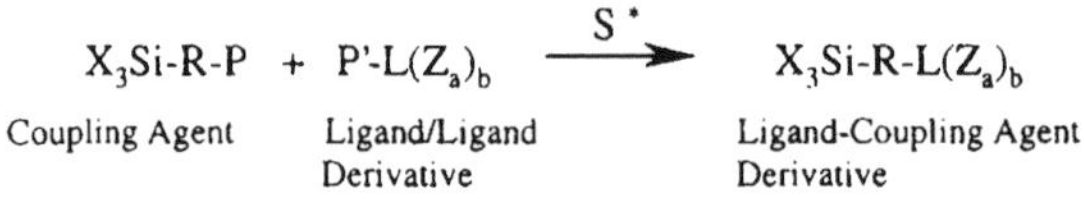

$$
X_3\text{Si-R-P} + \text{P'-L}(Z_a)_b \xrightarrow{\ S^{\,*}\ } X_3\text{Si-R-L}(Z_a)_b
$$

Coupling Agent   Ligand/Ligand   Ligand-Coupling Agent
                Derivative   Derivative

*: Different Reaction Schemes to Attach Ligand

Step 2: Immobilization of Ligand Coupling Agent Derivative

$$
\begin{bmatrix} \text{OH} \\ \text{OH} \\ \text{OH} \end{bmatrix} + X_3\text{Si-R-L}(Z_a)_b \longrightarrow \begin{bmatrix} \text{O} \\ \text{O} \\ \text{O} \end{bmatrix}\!\!\text{Si-R-L}(Z_a)_b + 3HX
$$

<u>Choice of Ligand Attachment Scheme:</u>
1. **Depends on Reactive Groups and Conditions**
2. **Desire to Achieve Ligand with Specific Donor Atoms and Preferred Geometry**
3. **Desire to Achieve High Ligand Density on the Support Surface**

Figure 7   Covalent attachment to support by method B.

the silica surface. Mattigod et al. [49] reported that theoretically up to 3.2 mmol/g of mercury uptake could be achieved by this method.

c. Organo-Ceramic Polymers Sol-gel chemistry provides a fundamentally different approach in the synthesis of organo-ceramic adsorbents. Sol-gel synthesis has been widely studied in glassmaking, optics, and coating processes [50,51]. Recently the incorporation of specific organic molecules in silica networks has been intensively studied for optical purposes [52,53]. This incorporation technique can be adopted in the synthesis of adsorbents. The basic chemistry is virtually identical and the concept is illustrated in Fig. 8. Advantages of these materials are high ligand densi-

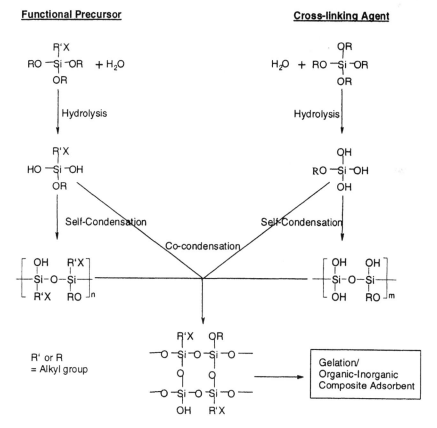

Figure 8   Sol-gel synthesis of an adsorbent. R′ and R are alkyl groups and X is the chelating molecule.

ties, homogeneous distribution of the functional moiety throughout the matrix, and controlled pore characteristics.

In our laboratory, a series of adsorbents are being synthesized through this sol-gel chemistry for the separation of heavy metals and noble metals from aqueous solution. For example, an adsorbent is synthesized with 3-mercaptopropyltrimethoxysilane and tetraethoxysilane as sol precursor and crosslinking agent, respectively. After optimization of the synthesis conditions—such as the composition of reaction mixture, types of solvent, concentration of catalyst used, and the hydrolysis/condensation time of both silanes—the adsorbent is shown to have high metal uptake capacity, selectivity, and stability as well as reasonable pore characteristics. The maximum uptake capacities of this adsorbent are 220 mg/g for cadmium and 1284 mg/g for mercury at pH 5–6.

For the synthesis condition optimization, NMR spectroscopy is one of the useful tools to identify the effect of each adjustable synthesis parameter on the adsorbents characteristics and consequently on metal uptake capacity and pore characteristics. The details of these studies are to be reported elsewhere.

III. CHARACTERIZATION OF EXTRACTION MATERIALS

The adsorbent materials are usually characterized for a variety of chemical and physical properties. Chemical properties of the materials include pH-dependent metal uptake, maximum metal uptake capacity, acid–base characteristics, metal selectivity, and chemical stability. Physical properties of the materials include mechanical strength and pore characteristics. A number of techniques discussed here are employed to determine these properties.

A. Chemical Properties of Solid Extractants

The metal extractions with solid adsorbents and ion exchangers occur through cationic and anionic adsorption/ion exchange mechanisms. The mechanisms are dependent on the acid–base characteristics of the functional ligands and the existing species of metal ions in the aqueous solutions.

Metal cations can react with a neutral ligand having a hydrogen bond by replacing the hydrogen ion, or with a negatively charged ligand as shown in the following equations:

Cation extraction:

$$M_{aq}^{z+} + n\overline{RH} \leftrightarrow \overline{R_nM^{(z-n)+}} + nH^+ \tag{1}$$

or

$$\overline{RH} \leftrightarrow \overline{R^-} + H^+ \tag{2}$$

$$M_{aq}^{z+} + n\overline{R^-} \leftrightarrow \overline{R_nM^{(z-n)+}} \tag{3}$$

Here M_{aq}^{z+} is a metal cation, $\overline{RH}$ is an unreacted neutral ligand on the solid surface, $\overline{R_nM^{(z-n)+}}$ is a metal–ligand complex on the solid surface, H^+ is a hydrogen ion, and $\overline{R^-}$ is a negatively charged ligand on the solid surface. For an example of Eq. (1), thiol ligands (—SH) react with cadmium ions and release hydrogen ions. It is known that the hydrogen replacement reaction is slow due to strong hydrogen bonding. For an example of Eqs. (2) and (3), carboxylic groups (COOH) can easily be deprotonated (COO⁻) at pH higher than 6 and can form complexes with cobalt.

The anionic metal adsorption process normally proceeds in two steps: first, protonation of the ligand and, second, anionic metal complex adsorption as shown in the following equations:

Anion extraction:

$$\overline{RN} + H^+ \leftrightarrow \overline{RNH^+} \tag{4}$$

$$M(A)_n^{z-} + m\overline{RNH^+} \leftrightarrow \overline{[(RNH)_mM(A)_n]^{(z-m)-}} \tag{5}$$

Here $\overline{RN}$ is an anionic extractant ligand on the solid surface, $\overline{RNH^+}$ is a protonated ligand on the solid surface, and $\overline{[(RNH)_mM(A)_n]^{(z-m)-}}$ is a metal ion–ligand complex on the solid surface.

For some cases, internal cation exchange and partial dissociation of the surface complex can occur, such as

$$\overline{[(RNH)_mM(A)_n]^{(z-m)-}} \leftrightarrow \overline{[(RN)_mM(A)_{n-m} \cdot m(HA)]^{(z-m)-}} \tag{6}$$

$$\overline{[(RN)_mM(A)_{n-m} \cdot m(HA)]^{(z-m)-}}$$
$$\leftrightarrow \overline{[(RN)_mM(A)_{n-m}]^{(z-m)-}} + m(HA) \tag{7}$$

In anionic metal ion extraction, the available functional ligand for the metal extraction is strongly dependent on the proton uptake of the extractant at a given pH. Hence, the metal uptake capacity of an anionic extractant is determined by both the proton uptake capacity of the extractant and the metal anionic speciation in the aqueous solution at a given pH.

1. pH Dependency of Extraction

The pH dependency of extraction is usually studied by conducting pH isotherm experiments. The pH isotherm provides useful information as it relates the extent of metal ion extraction with the solution pH. This information can be employed to determine conditions to extract and strip metal ions from the solid extractant. For example, cadmium extraction with thiol-based ligand adsorbent shows the highest uptake ability for pH higher than 5 and lowest metal capacity uptake capacity for pH below 2, as shown in Figure 9.

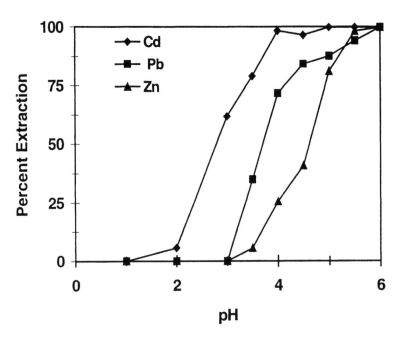

Figure 9 pH isotherm of cadmium, lead, and zinc on ICAA-S. Equilibration time = 1 h; weight of material = 1 g; volume of aqueous solution = 25 mL; initial metal concentrations = 200 ppm. (From Ref. 28.)

pH isotherms for anionic metal adsorption normally show higher metal uptake at lower pH values and lower uptake at higher pH values. However for some cases, such as chromium(VI) adsorption on anion adsorbents, as shown in Fig. 10, adsorption caps are observed. This is due to the formation of unfavorable aqueous metal species for the adsorption process at lower pH values.

2. Capacity

Theoretical uptake capacity of a solid extractant for a specific metal ion is the stoichiometric amount of the functional ligand on the solid surface. The theoretical uptake capacity can be determined by determining the moles of ligands per unit weight of the solids, the so-called ligand density. This measurement is possible by an elemental analysis or acid–base titration. For an example, Zhmud and Sonnefeld [54] synthesized organo-

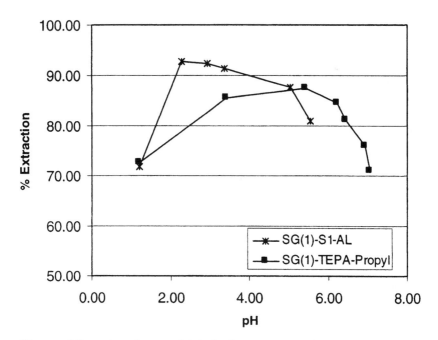

Figure 10   pH isotherms of Cr(VI) adsorption on SG(1)-TEPA-Propyl and SG(1)-S1-AL.

ceramic extractants by using aminopropyltriethoxylsilane and tetraethoxy-silane. They performed three different methods to determine the ligand densities of their materials: elemental analysis, thermal analysis, and acid–base titration. The results from the three different methods are compatible, as shown in Table 4. The details of the analysis can be found in their publication [54].

Actual maximum uptake capacities of the metal by the solid extractants can be determined either by batch experiments or by breakthrough experiments. In a batch system, a specified amount of the extractant is equilibrated with a concentrated metal solution for which complete saturation of the extractant is expected. Also, it is possible to determine the maximum metal uptake capacity by conducting a breakthrough curve experiment. This method is more practical than the batch method because the maximum metal uptake capacity can be determined at an operational condition. For example, an experiment was conducted to determine the breakthrough capacity of ICAA-S for cadmium ion adsorption, as shown in Fig. 11 [28]. The maximum cadmium uptake is reported as 71.1 mg/g at the conditions specified.

3. Selectivity

The metal selectivity of a specific solid extractant depends on the physicochemical property of the functional ligand incorporated in the solid

Table 4   The Amino Group Loadings (mmol g^{-1})

Method of determination	Matrix			
	APS-1	APS-2	MAPS-1	MAPS-2
Elemental analysis	3.1 ± 0.2	1.5 ± 0.1	1.5 ± 0.1	2.1 ± 0.2
Titration[a]	3.3 ± 0.3	1.0 ± 0.1	—	2.2 ± 0.2
Thermal analysis	3.5 ± 0.3	2.4 ± 0.2[b]	—	

[a] MAPS-1 proved to be hydrophobic, so it was not titrated.

[b] When there are two or more different groups being destroyed at near the same temperature, the thermal analysis only allows for the determination of the summed loading of all the groups present and not the distinguishing of the individual contribution of each of them.

Source: Ref. 54.

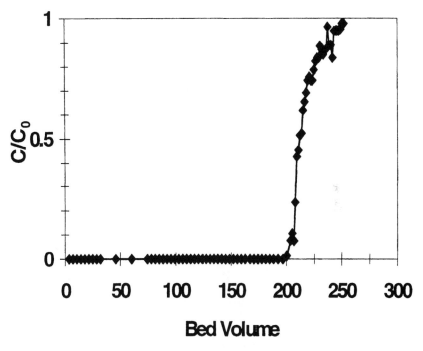

Figure 11   Laboratory scale breakthrough curve of cadmium on ICAA-S. C_0 = 197.5 ppm; pH = 6.3; weight of material = 3.97 g. Maximum cadmium uptake capacity = 71.1 mg/gm. (From Ref. 28.)

matrix, the oxidation state of metal ions, and the geometry of the surface. As mentioned earlier, usually a functional ligand that contains oxygen prefers to chelate with hard metal ions and sulfur-containing ligands easily form complexes with soft metal ions. However, the chelation between metal ions and the functional ligands on the solid surface may occur differently compared to the liquid–liquid system because of the complexity of the solid matrix and static hindrance effects. One can determine the selectivity of a solid extractant by equilibrating a unit mass of the solid extractant with a solution containing equimolar metal ions for a desired pH interval, such as shown in Fig. 12 [55].

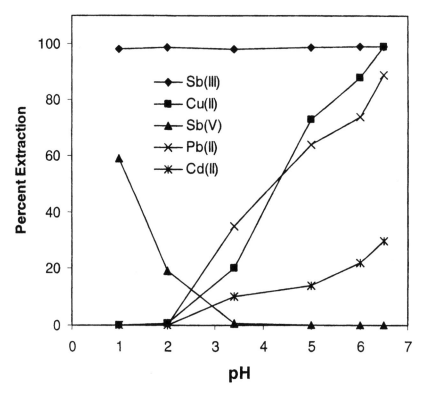

Figure 12   Effect of the pH of aqueous solutions on the adsorption of metal ions on ICAA-PPG. Equilibrium time = 1 h; weight of ICAA-PPG = 1 g; volume of aqueous solution = 25 mL; C_0 = 200 mg/L. (From Ref. 55.)

Selectivity of an adsorbent for various metal ions used to be expressed as ratios of distribution coefficients for the metal ions. For a binary system, the selectivity ($\alpha_{A/B}$) of an adsorbent for species A over species B can be represented as

$$\alpha_{A/B} = \frac{Kd_A}{Kd_B} \tag{8}$$

where Kd_i is the distribution coefficient for metal ion i, and is defined as the amount of adsorbed metal per unit mass of the adsorbent divided by the equilibrium metal concentration in the solution.

Generally, chelating ligands have higher selectivity for transition metal ions over alkaline metal ions. This fact is very important for environmental and hydrometallurgical applications of adsorbents. For example, a bench-scale breakthrough experiment with an organo-ceramic adsorbent, ICAA-S, was performed to show the selective removal of cadmium from a calcium-enriched solution [28], as shown in Fig. 13. In transition metal ions, chelating ligands have normally higher selectivity for metal ions that have higher valency than those with lower valency.

4. Stability
Stability of the solid extractants is very important because the number of cycles of the extractants that can be used is directly related to the life

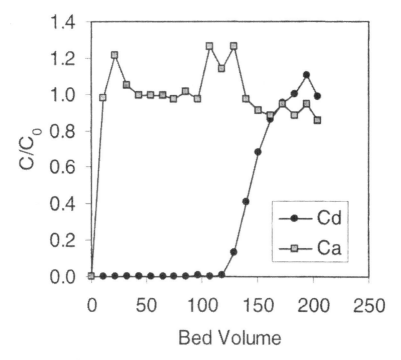

Figure 13   Selective removal of cadmium with ICAA-S bench-scale breakthrough curve. [Cd(II)] = 193.5 ppm; [Ca(II)] = 900 ppm; pH = 5.7. (From Ref. 28.)

cycle costs. Effects such as oxidation of ligands, irreversible metal adsorption on ligands, and cleavage of the ligand from the solid network, for example, reduce the metal uptake capacity of a solid extractant when the extractant is used repeatedly. One can determine the stability of the adsorbent by repeating metal loading and stripping cycles at desired conditions. For example, Cyanex 302–impregnated adsorbent, ISPE-302, was tested for cadmium breakthrough and stripping in a column repeatedly and showed approximately 71% loss of original capacity after 20 loading and stripping cycles [29], as shown in Fig. 14. However, it was reported [28] that ICAA-S synthesized by covalent attachment of thiol moiety on silica showed only 23% capacity loss (from 71.1 to 54 mg/g) after 20 cycles when a similar stability test was performed.

B. Physical Properties of Solid Extractants

In addition to having high ligand density and good stability, the solid extractant should have accessible pores and good mechanical strength. The mechanical strength of the solid extractants can be measured by standard methods used for polymer resins. The pore characteristics such as pore diameter, surface area, and pore size distribution can be measured by BET analysis with gas adsorption and by mercury porosimetry [56–58].

Rigid pore structures of hydrous oxides permit better permeability of metal ions into the matrix and facilitate introduction of active groups. However, it is noted that the immobilization of functional groups reduces pore diameters, pore volumes, and surface areas of the materials, as is shown in Table 5 [11] as an example. Hence, one should choose a proper precursor chain length to avoid significant size reduction of pore geometry and minimize slow diffusional mass transfer problems.

Relatively uniform pore size distribution and moderate pore diameters of the solid extractants (10–100 Å) are important to permit pore accessibility and ease diffusion of metal ions. For example, Mercier and Pinnavaia [24] synthesized a thiol moiety–grafted meso-porous silica adsorbent and studied the effects of pore characteristics of the modified silica on mercury uptake capacity. They showed that the modified meso-porous

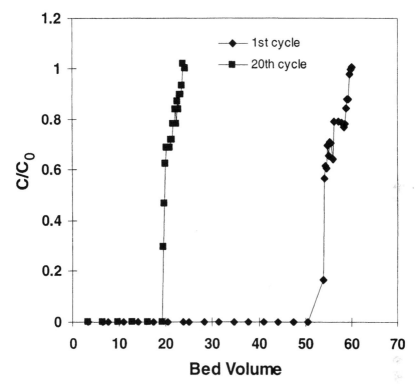

Figure 14 Breakthrough curves of cadmium on the ISPE-302 bed. Flow rate = 1.0–1.2 cm³/min; 1st cycle, [Cd²⁺] feed = 2.0 × 10⁻³ mol/dm³, pH = 6.5 (acetate buffer), uptake capacity = 0.175 × 10⁻³ mol/g; 20th cycle, [Cd²⁺] feed = 1.5 × 10⁻³ mol/dm³, pH = 6.5 (acetate buffer), uptake capacity = 0.5 × 10⁻⁴ mol/g. Plateau data after saturation are not shown. (From Ref. 29. Courtesy of the American Chemical Society.)

silica (MP-HMS-C8) that had smaller pore diameter (1.5 nm) had approximately 61% utilization of the total grafted SH sites for mercury uptake, but on the other hand the modified meso-porous silica (MP-HMS-C12) that had larger pore diameter (2.7 nm) showed 100% utilization of total SH. In addition, the mercury uptake rate was apparently slower when the average pore diameter decreased from 2.7 (MP-HMS-C12) to 2.0 nm (MP-MCM-41).

Table 5 Variation of Pore Structure Parameters of Silica by Means of Surface Modification

Type of packing	Specific surface area		Specific pore volume		Most frequent pore diameter D (nm)
	S_{BET} (m²/g)	$-\Delta S_{BET}$ (%)	V_p (mL/g)	$-\Delta V_p$ (%)	
Original SiO₂	300	—	1.32	—	12.6
Octylsilyl groups bonded	201	33	0.82	38	11.4
Octadecylsilyl groups bonded	139	54	0.52	61	7.0; 11.4

Source. Ref. 11.

IV. AQUEOUS PHASE EQUILIBRIUM

The adsorption of metal cations and metal anion species from aqueous solutions onto chelating ligands or ionic charged ligands on the support involves a surface reaction. Although the extent of adsorption can be described by a conventional isotherm expression such as the Langmuir or Fruendlich equations, it is more appropriate to consider ion adsorption with a model based on chemical reactions between functional ligands and metal ion species. To describe the adsorption phenomena with chemical reactions, one needs to understand the aqueous phase equilibria because metal ions can exist as cationic or anionic complexes depending on the oxidation state of the metal, the types of counterions present, the molar ratio of the metal ions to the existing counter anions, and solution pH. The existence of various metal species sometimes causes difficulty in describing the adsorption phenomena, especially in metal chelation reactions because the affinity of a functional ligand toward various metal ion species of a given metal is different. Hence, the consideration of aqueous phase equilibria will provide a better description of sorption processes.

In cationic metal systems, the metal cations exist at various valency states by forming complexes with aqueous anions or due to oxidation states. For example, Larson and Wiencek [59] studied mercury nitrate extraction with oleic acid as a liquid ion exchanger, wherein the aqueous phase equilibria of the mercury nitrate system was employed to model the uptake of cationic mercury species with the extractant. The developed model incorporating the aqueous phase equilibria could represent the experimental data well. Lee and Tavlarides [60] studied the extraction of iron(III) from sulfuric acid solutions with β-alkenyl-8-hydroxyquinoline extractant. They proposed a kinetic model including aqueous phase speciation of iron(III) in sulfate solution to describe the extraction kinetics.

In anionic systems, knowledge of speciation of metal complexes is more important than cationic systems to understand the adsorption phenomena. Quantitative existence of two or three anionic metal complexes in the aqueous solution makes it difficult to describe the adsorption of such complexes. For example, Sanjiv et al. [61] studied the adsorption of noble metal ions on a polyaniline adsorbent and showed that the distribution coefficients of noble metal ions were dependent on the types of counterions (sulfate, chloride), the counterion concentrations, and the solution pH. Similarly, Larson and Wiencek [59] showed the effects of chloride concentration and the solution pH on the extraction of mercury chloride with an anion liquid ion exchanger such as tri-isooctylamine. In their study, the speciation change of mercury chloride anions in the solution significantly affected extraction efficiency of the extractant. By considering the aqueous phase equilibria of mercury and the mercury extraction reaction with the reactant, they could predict the extraction of total mercury. Lee et al. [62] studied the adsorption of copper cyanide on an adsorbent, SG(1)-TEPA-propyl, and proposed an adsorption model which considered two types of copper cyanide complexes as adsorbates.

The computation of speciation of metal ions in aqueous solutions can be performed by considering mass balances of metal ion species in the solutions [63–67] along with known stability constants of the metal complexes [68] or by using available metal speciation software such as MINEQL [69]. Several traditional examples will be discussed here to illustrate the usefulness of the analysis to explain complex systems.

A. Cationic System: Iron(III) Sulfate Extraction

For the modeling of extraction kinetics of iron(III) from sulfuric acid with an extractant, β-alkenyl-8-hydroxyquinoline, an aqueous phase speciation model was developed to compute the concentrations of Fe(III) and proton [70]. Nine ionic equilibrium relationships, mass balances for total sulfur and iron concentrations, electroneutrality of the solution, and ionic strength are considered to solve the concentration of each constituent in the solution. The considered equilibrium relationships are

$$Fe_2(SO_4)_3 \xrightleftharpoons{k_1} 2Fe^{3+} + 3SO_4^{2-} \tag{9}$$

$$HSO_4^- \xrightleftharpoons{k_2} H^+ + SO_4^{2-} \tag{10}$$

$$Fe^{3+} + SO_4^{2-} \xrightleftharpoons{k_3} FeSO_4^+ \tag{11}$$

$$Fe^{3+} + HSO_4^- \xrightleftharpoons{k_4} FeHSO_4^{2+} \tag{12}$$

$$Fe^{3+} + 2SO_4^{2-} \xrightleftharpoons{k_5} Fe(SO_4)_2^- \tag{13}$$

$$Fe^{3+} + H_2O \overset{k_6}{\rightleftharpoons} FeOH^{2+} + H^+ \tag{14}$$

$$2FeOH^{2+} \overset{k_7}{\rightleftharpoons} Fe(OH)_2Fe^{4+} \tag{15}$$

$$Fe^{3+} + 2H_2O \overset{k_8}{\rightleftharpoons} Fe(OH)_2^+ + 2H^+ \tag{16}$$

$$H_2O \overset{k_9}{\rightleftharpoons} H^+ + OH^- \tag{17}$$

The mass balances, electroneutrality, and ionic strength relationship are as follows:

Sulfate balance:

$$[S]_T = [SO_4^{2-}] + [HSO_4^-] + [FeSO_4^+]$$
$$+ [FeHSO_4^{2+}] + 2[Fe(SO_4)_2^-] \tag{18}$$

Iron balance:

$$[Fe]_T = [Fe^{3+}] + [FeSO_4^+] + [FeHSO_4^{2+}] + [Fe(SO_4)_2^-]$$
$$+ [FeOH^{2+}] + 2[Fe(OH)_2Fe^{4+}] + [Fe(OH)_2^+] \tag{19}$$

Electroneutrality:

$$[H^+] + 3[Fe^{3+}] + [FeSO_4^+] + 2[FeHSO_4^{2+}] + 2[FeOH^{2+}]$$
$$+ 4[Fe(OH)_2\ Fe^{4+}] + [Fe(OH)_2^+] = 2[SO_4^{2-}]$$
$$+ [HSO_4^-] + [Fe(SO_4)_2^-] + [OH^-] \tag{20}$$

Ionic strength (I):

$$I = \frac{1}{2} \sum_{j=1}^{n} (z_j)^2 C_j \tag{21}$$

where z_j is the charge on the jth species, C_j is the concentration of the jth species, and n is the number of species in the solution.

The speciation of iron complexes can be computed by simultaneously solving mass balance equations, electroneutrality, and ionic strength equations in which the equilibrium relationships [Eqs. (9)–(17)] are incorporated. As a result of the computation, it is shown that various

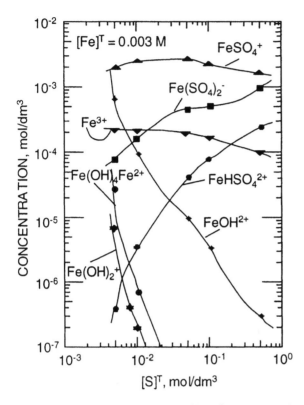

Figure 15 Concentration profiles of various iron ionic species as a function of total sulfur concentration at a constant total iron concentration of 0.003 mol/dm³. (From Ref. 70.)

types of iron sulfate and iron oxide complexes coexist in the sulfuric acid, as shown in Fig. 15.

These results were used to estimate $[Fe^{3+}]$ and $[H^+]$ to determine an interfacial kinetic model for the iron(III) sulfate–sulfuric acid, β-alkenyl-8-hydroxyquinoline-xylene extraction system [60].

B. Anionic System: Copper Cyanide Extraction

Copper cyanide exists mainly in the three different complex forms— $Cu(CN)_2^-$, $Cu(CN)_3^{2-}$, and $Cu(CN)_4^{3-}$ —in cyanide solutions, depending on the molar ratio of total copper to total cyanide and the solution pH. The equilibria of the aqueous copper cyanide solution can be represented by [71]

$$Cu^+ + CN^- \leftrightarrow CuCN(s) \tag{22}$$

$$Cu^+ + 2CN^- \leftrightarrow Cu(CN)_2^- \tag{23}$$

$$Cu^+ + 3CN^- \leftrightarrow Cu(CN)_3^{2-} \tag{24}$$

$$Cu^+ + 4CN^- \leftrightarrow Cu(CN)_4^{2-} \tag{25}$$

$$H^+ + CN^- \leftrightarrow HCN \tag{26}$$

$$H^+ + OH^- \leftrightarrow H_2O \tag{27}$$

To compute the species concentrations in the aqueous solution, the resulting equilibrium relations from Eqs. (22)–(27) are solved along with the mass balance for total copper ($[Cu]_T$) and total cyanide ($[CN]_T$) and a charge balance, as shown:

Total copper mass balance:

$$[Cu]_T = [Cu^+] + [Cu(CN)_2^-] + [Cu(CN)_3^{2-}]$$
$$+ [Cu(CN)_4^{3-}] \tag{28}$$

Total cyanide mass balance:

$$[CN]_T = [CN^-] + 2[Cu(CN)_2^-] + 3[Cu(CN)_3^{2-}]$$
$$+ 4[Cu(CN)_4^{3-}] + [HCN] \tag{29}$$

Charge balance of aqueous solution:

$$[Na^+] + [Cu^+] + [H^+] = [CN^-] + [Cu(CN)_2^-]$$
$$+ 2[Cu(CN)_3^{2-}] + 3[Cu(CN)_4^{3-}] + [OH^-] + [Cl^-] \tag{30}$$

The results of the computation of the aqueous phase speciation given in Fig. 16 show the variation of the copper cyanide species with respect to the solution pH and the ratio of total copper to total cyanide in the solution [62]. These results indicate that at lower pH, the cyanide solution equilibria favor the formation of hydrogen cyanide [HCN(aq)], which results in the decomposition of divalent and trivalent copper cyanide complexes to the lower valent complexes. The fraction of monovalent species

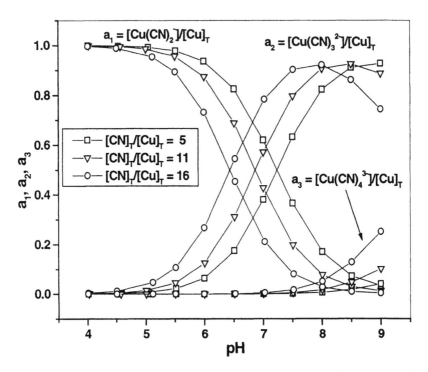

Figure 16 Computed speciation of copper cyanide complexes in cyanide solutions with respect to the solution pH.

decreases, while the fraction of divalent species increases, as pH increases. Concentrations of trivalent species are negligible below pH 7.5.

This analysis is employed to determine solution compositions for optimal sorption and to develop appropriate models to describe the adsorption process as described later.

V. METAL ION ADSORPTION ON SUPPORTED EXTRACTANTS

The interaction between metal ions and the functionalized organic molecules on the surface of adsorbents can be understood in parallel to that between metal ions and the molecules in liquid phase, although there may be geometrical restrictions on the solid adsorbents.

Many attempts to model the metal adsorption on activated carbons and metal oxides have been performed by considering the electrostatic interaction between the hydrolyzed surface sites and ions in the solution. Among these electrostatic models, some models interpret the adsorption phenomena by using electrical double layer (edl) theory, in which metal ions are closely packed due to the charge potential of the surface [72,73]. In addition to the consideration of edl, some models have been developed based on the surface complex formation between variously charged surface sites of the metal oxides and metal ions [74–77] such as shown below:

Ionization of hydroxyl groups of metal oxide surface [77]:

$$SOH + H^+ \leftrightarrow SOH_2^+ \tag{31}$$

$$SOH \leftrightarrow SO^- + H^+ \tag{32}$$

Metal and surface site complex formation:

$$SOH^{1/2-} + M^{2+} \leftrightarrow SOH^{1/2-} M^{2+} \tag{33}$$

However, the electrostatic model cannot represent the reactions between organic functional molecules and metal ions, such as ion exchange reactions and metal chelate formations. Many models have been proposed for the extraction of metal ions in solvent extraction, ion exchange, and adsorption with metal chelating ligands. These models usually consider the chemical reactions which include acid–base characteristics of the functional ligand and the coordination between the immobilized functional groups and metal ions. In addition, the aqueous phase speciation of the metal needs to be considered when there is a substantial amount of two or more metal species in the aqueous phase.

Normally, the functional groups on the support materials are activated to adsorb metal ions either by deprotonation or by protonation in aqueous solutions. Hence, the characteristics of acid–base properties of the adsorbents take an important role in the metal adsorption processes.

For example, the adsorption capacity of copper cyanide of an inorganic chemically active adsorbent, SG(1)-TEPA-propyl, as shown in Fig. 17, is determined by the degree of protonation of the functional groups at given pH values [62]. The pKa values of SG(1)-TEPA-proplyl were determined by potentiometric titration and by assuming the following mechanism:

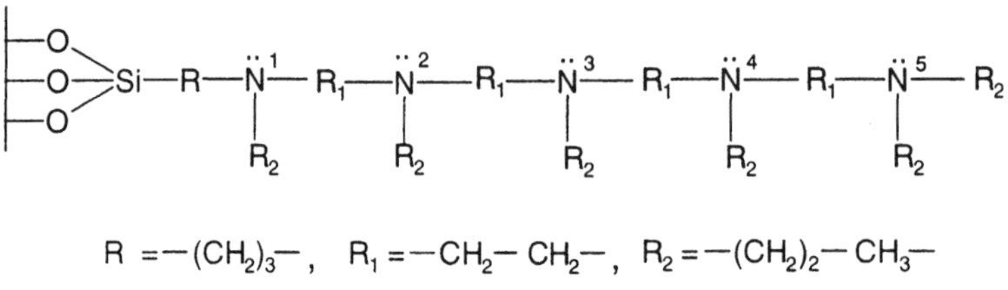

$$R = -(CH_2)_3-, \quad R_1 = -CH_2-CH_2-, \quad R_2 = -(CH_2)_2-CH_3-$$

Figure 17 Molecular structure of SG(1)-TEPA-propyl. (From Ref. 62. Courtesy of the American Chemical Society.)

$$\overline{RNH_i^+} \leftrightarrow \overline{RN}_i + H^+ \qquad Ka_i, i = 1, \ldots 5 \qquad (34)$$

where i refers to the ith site on a single ligand chain, $\overline{RN}$ is the unprotonated site, and $\overline{RNH^+}$ is the protonated site. The sum of the fraction of total protonated sites can be expressed as

$$\frac{[\overline{RNH^+}]_T}{[\overline{RN}]_{0,T}} = \frac{1}{5} \sum_{i=1}^{5} \frac{1}{10^{pH-pKa_i} + 1} \qquad (35)$$

where $[\overline{RN}]_{0,T}$ is total number of amine groups per unit mass of adsorbent and equals $5[\overline{RN}]_{0,T}$, and $[\overline{RNH^+}]_T$ is the total number of protonated amine groups per unit mass of adsorbent. The five pKa values are estimated by nonlinear parameter fitting for the titration data as shown in Fig. 18.

The incorporation of aqueous phase speciation in the modeling of metal ion uptake on functionalized adsorbents is frequently found in the literature. For example, Akita and Takeuchi [78] studied and modeled the adsorption of zinc chloride on a tri-n-octylamine–impregnated macromolecular resin. In their study, they proposed an adsorption mechanism, and the equilibrium constant for the reaction was computed by considering the aqueous phase zinc chloride speciation as shown:

Proposed adsorption mechanism:

$$\overline{(R_3N)}_{org} + HCl \leftrightarrow \overline{(R_3NH^+Cl^-)}_{org} \qquad (36)$$

$$2\overline{(R_3NH^+Cl^-)}_{org} + ZnCl_2 \leftrightarrow \overline{[(R_3NH^+)_2ZnCl_4^{2-}]}_{org} \qquad (37)$$

Aqueous phase speciation of zinc chloride in chloride solutions:

$$Zn^{2+} + iCl^- \leftrightarrow ZnCl_i^{2-i} \qquad i = 1, \ldots, 4 \qquad (38)$$

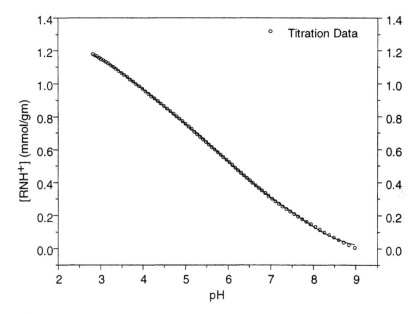

Figure 18 Titration curve of SG(1)-TEPA-propyl. Titrant = 0.1 M HCl; ionic strength = 0.1 (0.1 M NaCl); solid line = 5 pKa model with pKa$_1$ = 3.26, pKa$_2$ = 4.46, pKa$_3$ = 5.61, pKa$_4$ = 6.59, pKa$_5$ = 7.98, and [RN$_0$]$_T$ = 1.25 mmol/g. (From Ref. 62. Courtesy of the American Chemical Society.)

The aqueous phase speciation is correlated with the adsorption equilibrium equation through the mass balance of the zinc chloride in aqueous phase as shown:

$$[Zn]_{total} = [Zn^{2+}] + [ZnCl^+] + [ZnCl_2]$$
$$+ [ZnCl_3^-] + [ZnCl_4^{2-}] \quad (39)$$

In their model, the authors assumed only one metal species (ZnCl$_2$) was to be adsorbed on the adsorbent.

In a similar fashion, Kataoka and Yoshida [79] studied the adsorption of mercury chloride on a chloride form anion exchanger. In their model, the authors considered the mercury complex on the adsorbent surface as well as the speciation of mercury in the aqueous phase simultaneously. The considered mechanism of aqueous phase equilibria of mercury chloride complex is

$$2HgCl_2 + H_2O \leftrightarrow (HgCl)_2O + 2H^+ + 2Cl^- \tag{40}$$

$$HgCl_2 + Cl^- \leftrightarrow HgCl_3^- \tag{41}$$

$$HgCl_3^- + Cl^- \leftrightarrow HgCl_4^{2-} \tag{42}$$

The speciation in the solution is computed by considering the mass balances of total mercury and total chlorides, and the electroneutrality of the solution along with the proposed equilibria mechanism.

The proposed mechanism of ligand–mercury complexation in the solid phase is

$$\overline{R \cdot Cl^-} + HgCl_2 \leftrightarrow \overline{R \cdot HgCl_3^-} \tag{43}$$

$$\overline{2R \cdot Cl^-} + HgCl_2 \leftrightarrow \overline{R \cdot HgCl_4^{2-}} \tag{44}$$

$$\overline{R \cdot HgCl_3^-} + HgCl_2 \leftrightarrow \overline{R \cdot Hg_2Cl_5^-} \tag{45}$$

The speciation of ligand–mercury complexes is computed by considering the total mercury balance on the solid phase along with the ligand–mercury complexation equilibria.

Lee et al. [62] modeled the adsorption of copper cyanide on an organo-ceramic adsorbent, SG(1)-TEPA-propyl, by using a modified Langmuir isotherm. They assumed two major copper cyanide species were adsorbed on protonated ligands through the following mechanism:

$$\overline{RN} + H^+ \leftrightarrow \overline{RNH^+} \qquad K_a \tag{46}$$

$$A^- + \overline{RNH^+} \leftrightarrow \overline{RNHA} \qquad K_1 \tag{47}$$

$$B^{2-} + \overline{RNH^+} \leftrightarrow \overline{RNHB^-} \qquad K_2 \tag{48}$$

The concentrations of copper cyanide species in the aqueous phase were calculated by employing the aqueous phase equilibria as discussed in the previous section. The proposed isotherm is

$$q = f \frac{K_1 C_A + K_2 C_B}{1 + K_1 C_A + K_2 C_B} [\overline{RNH^+}]_T \tag{49}$$

Here, A^- is $Cu(CN)_2^-$ and B^{2-} is $Cu(CN)_3^{2-}$; $[\overline{RNH^+}]_T$ is the total moles of protonated sites per unit gram of the adsorbent and depends on pH as shown in Eq. (35); C_i is the equilibrium concentration of species i; and f is a correlation factor. It is observed that the copper cyanide complexes cannot be adsorbed to the theoretical maximum due to geometrical constraints. The correlation factor f accounts for this limitation, and the experimental value obtained is consistent with the molecular geometry of the complex and the atomic spacing between donor atoms on the ligand. The suggested adsorption model shows good agreement with the experimental data, as shown in Fig. 19.

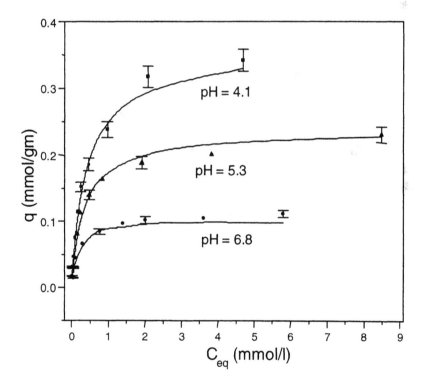

Figure 19   Adsorption isotherms of copper cyanide complexes at three different pH values. Symbols represent experimental data and solid lines represent the model, Eq. (49) with the values of $f = 0.4 \pm 0.01$, $K_1 = 2.43 \pm 0.15$, and $K_2 = 3.95 \pm 0.82$. (From Ref. 62. Courtesy of the American Chemical Society.)

VI. ADSORPTION KINETICS OF METAL ION ON ADSORBENTS AND ION EXCHANGERS

The adsorption of metal ions on adsorbents and ion exchangers is involved in complex chemical and physical phenomena. It is assumed that adsorption of metal ions takes place through three sequential steps: transport of the metal ions from bulk aqueous phase to the outer surface of adsorbents, diffusive transport (pore and surface diffusions) of metal ions from the pore solution to the functional sites on the adsorbent surface, and the chemical reaction between functional groups and metal ions on the adsorbent surface. Consideration of all three steps in the adsorption kinetics modeling is not an easy task and may not be necessary because either one or two of the three steps normally controls the overall adsorption rate. Hence, we will limit this discussion to adsorption kinetics of the transport rate–limited and chemical reaction rate–limited cases for illustrative purposes.

A. Mass Transfer–Limited Adsorption Kinetics

Numerous studies have discussed the adsorption kinetics of organic compounds [80,81] and gases [82] on various adsorbents and ion exchangers, and most of the studies dealt with the case of transport-limited adsorption kinetics. Assuming the transport-limited kinetics is true in most cases of metal adsorption processes, analyses have been successfully developed to model metal ion adsorption on hydrous oxides [83], activated carbon, and simple ion exchangers [84].

Generally, transport-limited kinetics are modeled by considering interphase and intraparticle transfer simultaneously. The interphase mass transfer relates the adsorption rate to the difference of adsorbate concentrations in the bulk phase and at the solid–liquid interface such as [80]

$$\frac{\partial q}{\partial t} = k_f a(C_b - C_s)$$ (50)

$$= 3 \frac{k_f}{a_p \rho_p}(C_b - C_s)$$

where q is average adsorbed phase concentration per unit mass, C_b and C_s are adsorbate concentrations in the bulk solution and at the interface of the solution and the solid particle, a_p is the particle radius, ρ_p is the particle density, a is outer surface area per unit mass of the particles, and k_f is external mass transfer coefficient.

Intraparticle mass transfer can be described with diffusion concepts of either pore diffusion or surface diffusion. Alternatively, one may use an effective diffusivity in which both diffusion coefficients are combined as one parameter. The diffusive transports are expressed as macroscopic mass balances in the solid phase such as [80]:

Pore diffusion:

$$\varepsilon_p \frac{\partial c}{\partial t} + \rho_p \frac{\partial q}{\partial t} = \frac{1}{r^2} \frac{\partial}{\partial r}\left(D_p r^2 \frac{\partial c}{\partial r}\right) \qquad 0 \le r \le a_p \qquad (51)$$

Surface diffusion:

$$\frac{\varepsilon_p}{\rho_p} \frac{\partial c}{\partial t} + \frac{\partial q}{\partial t} = \frac{1}{r^2} \frac{\partial}{\partial r}\left(D_s r^2 \frac{\partial q}{\partial r}\right) \qquad 0 \le r \le a_p \qquad (52)$$

Combined pore and surface diffusion:

$$\varepsilon_p \frac{\partial c}{\partial t} + \rho_p \frac{\partial q}{\partial t} = \frac{\rho_p}{r^2} \frac{\partial}{\partial r}\left(D_e r^2 \frac{\partial q}{\partial r}\right) \qquad 0 \le r \le a_p \qquad (53)$$

where c and q are the adsorbate concentration in the pore fluid and on the adsorbent, ε_p is the particle porosity, D_p, and D_s are the pore diffusivity and surface diffusivity, D_e is the effective diffusivity, which is defined as $D_e = [D_p/f'(C)](1/\rho_p) + D_s$, and $f'(c)$ is the derivative of q (adsorption isotherm expression) with respect to c such as $q = f(c)$ and $dq = f'(c)$ dc.

One of these diffusive mass transport equations can be solved numerically along with the interphase mass transfer equation and proper initial and boundary conditions. Both the interphase mass transfer coefficient and the diffusivities can be found from reported correlations, independent experiments, or least-square fitting of the adsorption kinetic data.

For example, Kim et al. [85] studied the adsorption of radionuclides from aqueous solutions by inorganic adsorbents and modeled a batch adsorption process by using the surface diffusion model. The authors assumed the first term in Eq. (52) was negligible compared to the second term and used the following initial and boundary conditions:

$$q(r,0) = 0 \tag{54}$$

$$\left.\frac{\partial q}{\partial r}\right|_{r=0} = 0 \tag{55}$$

$$\left.D_s \frac{\partial q}{\partial r}\right|_{r=a_p} = k_f(C_b - C^*) \tag{56}$$

For the local adsorption equilibrium, the Dubinin-Astakov (DA) adsorption isotherm was employed such that

$$C^* = C_s \exp\left[\left(-\ln\frac{q|_{r=a_p}}{q_m}\right)^{1/n}\left(\frac{E}{kT}\right)\right]^{-1} \tag{57}$$

where q_m is the amount adsorbed for monolayer formation on the adsorbent, E is the characteristic adsorption energy, and k is the Boltzmann constant. The adsorbate concentration change in the bulk phase was expressed by

$$\frac{dc}{dt} = -\frac{w}{v}D_s\frac{3}{a_p}\left.\frac{\partial q}{\partial r}\right|_{r=a_p} \tag{58}$$

where w is weight of the adsorbent and v is volume of the solution used.

The set of equations was nondimensionalized, transformed to a set of ordinary differential equations by using the orthogonal collocation method, and solved numerically. The model was successful to predict the rate of adsorption of the radionuclides as compared to experimental results. Many similar approaches using the surface diffusion model are reported in literature [26,86,87].

Also, examples using the pore diffusion model applied to describe metal adsorption kinetics exist in literature. The mathematical approaches are similar to the above, and details are in monographs [77,80].

B. Chemical Reaction Rate–Limited Adsorption Kinetics

Unlike the mass transfer–limited cases, less attention has been given to surface reaction–controlled adsorption processes, even though this limiting case is important in metal adsorption kinetics. Particularly, it is reported that chelation reactions of metal ions with some functional groups are slow [88]. Also, the reaction rates of metal chelation with one functional group vary with different metal ions and solution conditions, such as ionic strength and pH of the solution [88,89]. For example, Alexandratos et al. [90] compared the percent complexations of phosphonic acid immobilized resins after 30 min and 24 h of contact with various metal ions. As shown in Table 6, the gel-type resin forms 95.46 and 99.63% complexation with Eu(III), while 9.98 and 43.34% with Fe(III) after 30 min and 24 h contact time, respectively. This reveals the complexation rate of phosphonic acid with Eu(III) is much faster than that with Fe(III).

In another study, it is reported that the change of solution pH exerts a shift of adsorption chemistry and, consequently, affects the surface chemical reaction rates. Warshawsky et al. [88] studied the adsorption of platinum group metals on isothiouronium resin and observed that the adsorption rate of palladium chloride on the resin from 4.0 M HCl is faster than from 0.05 M HCl solution. They explained that the slow uptake after fast initial uptake in a 0.05 M HCl solution is due to slow coordination between palladium chloride and thiourea ligands.

Hence, adsorption kinetics limited by surface chemical reactions needs to be modeled for systems similar to the above by employing proper

Table 6 Comparison of Percent Complexation of Various Metal Ions on Phosphonic Acid–Impregnated Gel and MR-Type Polymer Resins after 30 min and 24 h of Contact Time

Polymer	Reaction time	Eu(III)	Fe(III)	Cu(II)	Pb(II)
Gel	30 min	95.46	9.98	39.07	86.71
Gel	24 h	99.63	43.34	52.82	90.58
MR	30 min	95.56	62.41	48.64	88.00
MR	24 h	100.00	98.96	52.22	88.00

Source: Ref. 90.

chemical reaction mechanisms. The chemical reactions between metal ions and the functional groups on adsorbent surfaces can be assumed the same as those for liquid–liquid systems. The chemical kinetics for the reaction of metal ions and functional ligands have been widely studied for many purposes, especially in solvent extraction processes, as stated earlier, and are well established. General chemical reaction schemes involved in adsorption processes can be found elsewhere [77].

Yiacoumi and Tien [91] suggested an adsorption kinetics model in which the surface–metal complexation reaction is the rate-limiting step. Even though this model was applied for the metal adsorption on hydrous oxides and the electrostatic effects were considered, this study provides insight for the modeling of chemical reaction rate–limited adsorption processes.

Also, there is an example of adsorption kinetics study that considers chemical reaction with transport phenomena simultaneously given by Lee et al. [92]. They studied the adsorption of copper ions on a LIX-84–impregnated silica adsorbent. To model the adsorption process, they incorporated the chemical kinetics of copper sulfate extraction with the LIX-84 for the liquid–liquid system into their pore diffusion model.

VII. METAL ION ADSORPTION IN PACKED BEDS

Solid supported extractants can be employed in fixed-bed contactors to extract metal ions from solutions. In this case the following mass transfer processes may be present: interpellet mass transfer, which refers to the diffusion and mixing of metal ion in fluid occupying the spaces between pellets; interphase mass transfer, which is the transfer of metal ion across the fluid pellet interface; and intraparticle mass transfer, which refers to the diffusion of metal ions in the pellet (either in the pore solution or adsorbed on the pore surface). A schematic for a packed bed adsorption process is shown in Fig. 20.

To analyze or to predict the metal adsorption in a packed column, we need to consider these mass transfer phenomena and the chemical reaction of the metal adsorption. We discussed the interphase and intra-

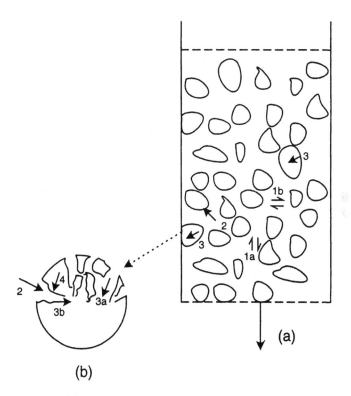

Figure 20   Mass transfer in adsorption processes: (a) fixed beds and (b) intrapellet mass transfer. 1a, axial dispersion; 1b, radial dispersion; 2 interphase mass transfer; 3, interpellet mass transfer; 3a, pore diffusion; 3b, surface diffusion; 4, adsorption. (From Ref. 80.)

phase mass transfer and the chemical kinetics in the previous chapter. The expressions for these adsorption steps are also valid for the adsorption analysis in packed column operations. Hence, a mass balance equation for the interpellet mass transfer is additionally required to model the adsorption in packed columns.

A. Interpellet Mass Transfer

In fixed-bed adsorption, diffusion and mixing of metal ions in fluids occur due to adsorbate concentration gradients and nonuniformity of fluid flow.

This gives rise to axial dispersion in the main direction of fluid flow and radial dispersion in the direction transverse to the main flow. The former is usually undesirable since it reduces separation efficiency, whereas the latter is desirable as it equalizes concentrations at the same axial position and reduces axial dispersion. For the simple case of single-phase flow in a packed bed of cylindrical configuration, the conservation equation for a metal ion in the solution is [80]

$$\frac{\partial C_i}{\partial t} = \frac{E_r}{r}\frac{\partial}{\partial r}\left(r\frac{\partial C_i}{\partial r}\right) + E_z\frac{\partial^2 C_i}{\partial z^2} - \left(\frac{u_s}{\varepsilon}\right)\frac{\partial C_i}{\partial z} \tag{59}$$

where C_i = concentration of metal ion species i
   t, r, z = independent variables of time, radial distance, and axial distance, respectively
   u_s = superficial velocity
   ε = fixed-bed porosity
   E_r, E_z = radial and axial dispersion coefficients.

A great deal of effort has been expended to relate E_r and E_z to the operation and physical parameters and will not be discussed here.

The equations for mass transfer and chemical reaction on adsorbents can be solved along with appropriate assumptions to describe the behavior of a fixed-bed solid extractant column. In most cases, numerical computations are required to solve the set of partial equations except for some restricted cases, e.g., the assumption of a linear adsorption isotherm. Numerous examples can be found in the literature [80] and a few simple and relevant systems will be discussed here.

1. Case I: No Mass Transfer Limit

The extraction of copper cyanide species using SG(1)-TEPA-propyl [62] discussed in Section V was studied in a fixed-bed column filled with 75–270 μm particles. It was observed that the extraction rate of copper cyanide complexes on SG(1)-TEPA-propyl was too rapid to measure due to fast kinetic and pore diffusion (small particles). Accordingly, the adsorption behavior can be easily modeled using the equilibrium model in Section V and neglecting the interphase and intraparticle mass transfer phenomena. Restating Eq. (49),

$$q = f \frac{K_1 C_A + K_2 C_B}{1 + K_1 C_A + K_2 C_B} [\overline{RNH^+}]_T \tag{60}$$

Also, Eq. (51) can be rewritten by assuming no pore and surface diffusion limit:

$$\varepsilon_p \frac{\partial c_i}{\partial t} + \rho_p \frac{\partial q_i}{\partial t} = 0 \qquad 0 \le r \le a_p \tag{61}$$

Under the assumptions of no radial and axial dispersion ($E_r = E_z = 0$) and no interphase mass transfer resistance ($C_i = c_i$), Eqs. (59) and (61) can be combined and expressed as

$$\rho_b \frac{\partial q_i}{\partial t} = \left(\frac{u_s}{\varepsilon}\right) \frac{\partial c_i}{\partial z} \tag{62}$$

where ρ_b is a bulk density of the packed column.

Equations (62) and (60) are combined by using the first derivative of the Eq. (43) with respect to time and nondimensionalized to give

$$\frac{\partial C^*}{\partial z^*} + \frac{K}{(1 + KC^*)^2} \frac{\partial C^*}{\partial \theta^*} = 0 \tag{63}$$

$$C^*(0,\theta^*) = 1 \qquad C^*(z^*,0) = 0 \tag{64}$$

where

$$K = [K_1 \alpha_1 + K_2(1 - \alpha_1)] C_0$$

$$C^* = \frac{C_{aq}}{C_0}$$

$$z^* = \frac{z}{L}$$

$$\theta^* = \frac{\theta u_s}{L} \frac{C_0}{\rho_b f [RNH^+]_T}$$

$$\theta = t - z\left(\frac{\varepsilon}{u_s}\right),$$

C_0 is the feed concentration of total copper, C_{aq} is total copper concentration in the solution, L is the height of the packed column and α_1 is $[Cu(CN)_2^-]/C_{aq}$.

The differential equation with the given boundary conditions is solved numerically using a finite difference method and the computed effluent concentration of copper cyanide is compared with experimental data as shown in Fig. 21.

The developed model assuming no mass transfer limitations and equilibrium between solid and liquid phases represents the experimental breakthrough curve well. The experimental result shows that the slope of the curve at the breakthrough point is a sharp steplike response, which means there are negligible or no film resistances at the particle surface,

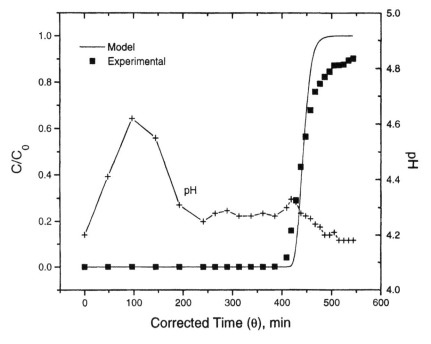

Figure 21 Breakthrough curve of copper cyanide complexes on SG(1)-TEPA-propyl at pH = 4.1. (From Ref. 62. Courtesy of the American Chemical Society.)

fast diffusion of copper cyanide complexes through the pores, and fast adsorption reaction on the functional sites.

2. Case II: Mass Transfer Limit and Linear Adsorption Isotherm

An exact solution and an approximate analytical solution for the modeling of adsorption processes in packed columns were developed by Rosen [93] for the mass transfer–limited and linear adsorption isotherm–governed cases. Additional assumptions were made to develop the analytical solution such as surface diffusion, no axial and radial dispersion effects, and constant superficial velocity through the column. The sets of governing equations are as follows:

Linear form of local adsorption isotherm:

$$q = Kc \tag{65}$$

Interpellet mass transfer, no axial and radial dispersion:

$$u_s \frac{\partial c_b}{\partial z} + \rho_b \frac{\partial \bar{q}}{\partial \theta} = 0 \tag{66}$$

Intrapellet mass transfer:

$$\left(\frac{\varepsilon_p}{\rho_p}\right) \frac{\partial c}{\partial \theta} + \frac{\partial q}{\partial \theta} = \frac{1}{r^2} \frac{\partial}{\partial r}\left(D_s r^2 \frac{\partial q}{\partial r}\right) \tag{67}$$

Interphase mass transfer:

$$\rho_p \frac{\partial \bar{q}}{\partial \theta} = \frac{3 k_f}{a_p}(c_b - c_s) \tag{68}$$

Initial and boundary conditions:

$$q = 0, \qquad \theta < 0$$

$$\frac{\partial q}{\partial r} = 0, \qquad r = 0$$

$$k_f(c_b - c_s) = D_s \rho_p \left(\frac{\partial q}{\partial r}\right) \qquad r = a_p$$

$$c_b = 0, \qquad z > 0, \theta < 0$$

$$c_b = c_{in}, \qquad z = 0, \theta > 0$$

where K is a distribution coefficient, k_f is a mass transfer coefficient, D_s is a surface diffusion coefficient, a_p is the particle radius, c_b is the adsorbate concentration in the bulk solution, c_s is the adsorbate concentration at external surface of the particle, c_{in} is the adsorbate concentration in feed solution, ε_p is a porocity of the adsorbent, ρ_p is a particle density, ρ_b is a bulk density, $\bar{q}$ is an average concentration on an adsorbent particle (defined as $3/a_p^3 \int_0^{a_p} qr^2 dr$), θ is a correct time [defined as $t - z(\varepsilon/u_s)$] t is the time scale, z is an axial coordinate of the column, ε is a bed porosity, and r is a radial coordinate of the particle.

The set of equations was solved by using the Laplace transform method. The exact solution was given as a complex infinite integral equation (see Rosen's paper [93] for details). In addition to the exact solution, Rosen provided an approximate solution for a limited case as follows:

$$\frac{c}{c_{in}} = \frac{1}{2}\left\{1 + \text{erf}\left[\frac{(3\sigma\theta/2) - \gamma x}{(\gamma x/5)^{1/2}}\right]\right\}$$

for the case of $\gamma x \geq 50$ and $\nu < 0.01$ (69)

where

$$\sigma\theta = 2D\frac{t - z\varepsilon/u_s}{a_p^2} \qquad \gamma x = 3D\frac{Kz\rho_b}{a_p^2 u_s}$$

$$\nu = \frac{DK\rho_p}{k_f a_p} \qquad D = \frac{D_s}{1 + \rho_p/\varepsilon_p K}$$

It was reported that the approximate solution could represent the breakthrough curve of strontium removal with zeolite [94], as shown in Fig. 22.

3. Case III: Mass Transfer Limit and Nonlinear Isotherm

The modeling of adsorption processes in packed columns for the case of mass transfer–limited adsorption with a nonlinear isotherm prohibits analytical solutions. Modeling for this case uses almost identical governing equations discussed in Case II except for the adsorption isotherm (mostly Langmuir, Freundlich, and modified Langmuir [84,95] isotherms for this case). Numerical computation techniques applied to solve the set of partial

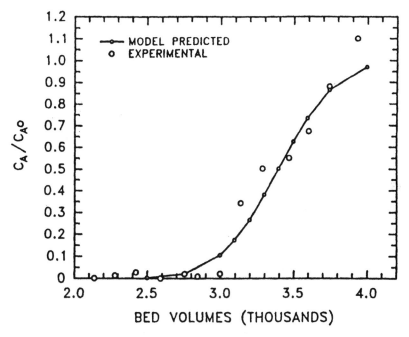

Figure 22   Model-predicted breakthrough curve. Flow rate = 0.25 L/min. (From Ref. 94.)

differential equations use either finite difference or orthogonal collocation methods. The details for these methods can be found elsewhere [96,97] and will not be discussed here.

VIII.   APPLICATIONS

Organo-ceramic extractants have multiple uses. Analytical applications include packing for a stationary phase as in HPLC or as a material for preconcentration of metal ions for metal analysis [14,35]. Industrial applications include hydrometallurgical separations, treatments of solutions in microelectronics fabrications, and purification of solutions in electrorefining. Examples of environmental separations include acid mine drainage remediation, trace metals removal from scrubbing solutions of flue gases of power plants, cleanup of streams from electroplating industries, and arsenic removal from aqueous runoffs. Several potential applications are given below to conclude this chapter.

A. Antimony Separation from Copper Electrorefining Solutions

Antimony and its compounds are listed among the most toxic elements of priority pollutants by the U.S. Environmental Protection Agency. Antimony exists in several industrial and mining wastes, such as chemical and allied products, glass products, electrical and electronic equipment, lead acid storage batteries, and copper electrorefining solutions.

The solid extractant immobilized with pyrogallol moiety (ICAA-PPG) shown in Fig. 23 has been applied for the separation of antimony from copper electrorefining solutions [55]. The ICAA-PPG has 43.1 mg/g antimony saturation capacity at pH 6. This extractant can selectively remove antimony from the aqueous solution containing other metal ions at pH less than 2, as shown in Fig. 12. Also, antimony can be selectively separated after loading with other metal ions on the extractant by sequential stripping with different solutions, as shown in Fig. 24.

Further details of this application can be found elsewhere [55].

B. Zinc, Cadmium, and Lead Separation

For the removal of zinc, cadmium, and lead from aqueous waste streams, an organo-ceramic extractant was synthesized in our laboratory by the method of ligand deposition on hydrophobized silica gels [29]. The functional ligand deposited on the silica is bis(2,4,4-trimethylpentyl)monothiophosphinic acid (Cyanex-302, American Cyanamide Co.). The synthesized material showed the selectivity series of $Cd^{2+} > Pb^{2+} > Zn^{2+}$.

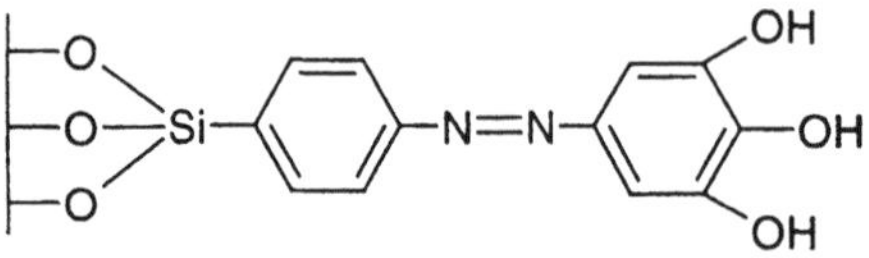

Figure 23 Molecular structure of bonded pyrogallol (ICAA-PPG).

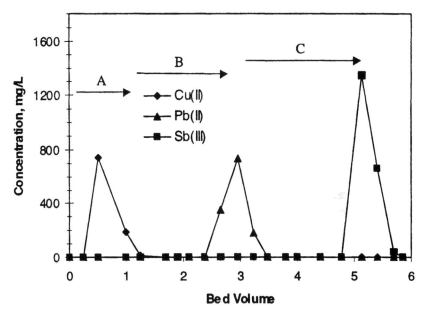

Figure 24   Separation of copper, lead, and antimony (III) by selective stripping from ICAA-PPG bed. Stripping solution A = 0.5 M sulfuric acid; stripping solution B = 0.5 M hydrochloric acid; stripping solution C = 4 M hydrochloric acid with 0.05 M potassium hydrogen tartrate. (From Ref. 55.)

A breakthrough curve experiment for the solution containing Cd^{2+}, Pb^{2+}, and Zn^{2+} ions was performed to show the selective extraction of metal ions, as shown in Fig. 25. The breakthrough of zinc ion occurred at 61 bed volumes, although both lead and cadmium continue to be removed to less than 8.0×10^{-6} mol/dm³. After 70 bed volumes the concentration of zinc in the effluent is greater than the feed concentration, while lead and cadmium are being adsorbed. The increase of the zinc concentration in the effluent is due to displacement of zinc by the Cd^{2+} and Pb^{2+} ions, which have higher affinities for the active sites than zinc. Similar behavior for displacement of adsorbed lead after 104 bed volumes shows that cadmium has the greatest affinity for this adsorbent.

The adsorbed metals can be recovered by selective stripping similar to the previous example for antimony (III) separation. Here, ad-

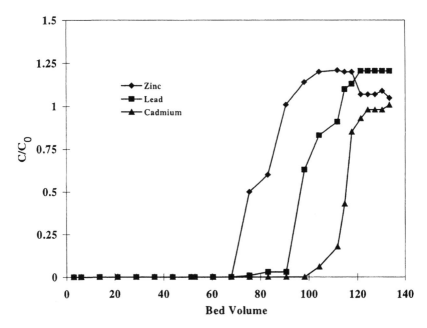

Figure 25 Simultaneous removal of zinc, lead, and cadmium with ICAA-302. Flow rate = 1.0–1.2 cm^3/min; pH = 6.3; [Zn^{2+}] = 55 mg/L; [Cd^{2+}] = 48 mg/L; [Pb^{2+}] = 44 mg/L. (From Ref. 29. Courtesy of the American Chemical Society.) 406. Copyright 1997 Am. Chem. Soc.)

sorbed lead, zinc and cadmium is shown to be selectively stripped using 0.1 M nitric acid, 1.0 M nitric acid, and 2.0 M hydrochloric acid respectively.

Further details of this application can be found elsewhere [29].

C. Berkeley Fit Problem

The Berkeley Pit is a vast open-pit mine located in an ore-rich section of southwestern Montana which was closed in the 1980s and has been targeted by the U.S. Department of Energy as a"demonstration site" for environmental reclamation [98]. The Berkeley Pit is filled with some 17 billion gallons of water which contains various types of metal ions, including iron, copper, zinc, manganese, magnesium, and aluminum at considerably high concentrations. A typical composition of the Berkeley Pit water is shown in Table 7. Metal separation from the solution is necessary from an environmental aspect and also one may recover useful metals economically. A process has been proposed to separate these metals by using various organo-ceramic adsorbents prepared in our laboratory.

One proposed scheme is to use three different adsorbents shown in Table 8 for a train of four columns for the separation of the major components. A schematic of the process is shown in Fig. 26. Here column A selectively removes Fe(III), as shown in Table 8. The second column with ICAA-C selectively removes Cu(II) and some Fe(II). In order to separate Fe(II) and Cd(II), Pb(II), and Zn(II), the third column with ICAA-D will remove Fe(II) preferentially over the other three metals. We have suggested other adsorbents for Fe(II) removal steps, but have not developed them at present. The final column, also with ICAA-D can remove Cd(II), Pb(II), and Zn(II). A major additional benefit with ICAA-D is

Table 7 Target Metals and Other Constituents in Berkeley Pit Water

Ions	Concentration (mg/L)	Other constituents	Value/ concentration
Aluminum	2.60	pH	2.85
Cadmium	2.14	$[Fe^{3+}]/[Fe]$	0.15
Calcium	456		
Copper	172		
Iron	1068		
Lead	0.031		
Magnesium	409		
Manganese	185		
Sodium	76.5		
Zinc	550		
Nitrate	<1.0 (as N)		
Sulfate	7600		

Source: Ref. 99.

Table 8 Adsorption of Metal Ions from Simulated Berkeley Pit Water

ICAAs material	pH	Percent extraction					
		Fe(III)	Cu(II)	Zn(II)	Cd(II)	Pb(II)	Fe(II)
ICAA-A	2.5–2.7	98.7[a]	6.4[a]	ND[a]	ND[a]	ND[a]	5.1[c]
		NA[b]	NA[b]	5.6[b]	5.6[b]	34.0[b]	NA[b]
ICAA-C	2.5–2.7	43.7[a]	93.1[a]	ND[a]	ND[a]	ND[a]	11.7[c]
		NA[b]	NA[b]	ND[b]	ND[b]	ND[b]	NA[b]
ICAA-D	2.5–2.7	NA[b]	NA[b]	99.6[b]	99.8[b]	99.7[b]	98.7[c]

[a] Feed solution composition: [Fe(III)] = 200 ppm, [Cu(II)] = 200 ppm, [Zn(II)] = 200 ppm, [Cd(II)] = 3.8 ppm, [Pb(II)] = 4.1 ppm, [SO$_4^{2-}$] = 8000 ppm, pH = 2.5.
[b] Feed solution composition: [Zn(II)] = 200 ppm, [Cd(II)] = 3.8 ppm, [Pb(II)] = 4.1 ppm, [SO$_4^{2-}$] = 8000 ppm, pH = 2.5.
[c] [Fe(II)] = 200 ppm, [SO$_4^{2-}$] = 8000 ppm, pH = 2.5.
ND: not detectable. NA: not applicable.
Source. Ref. 98.

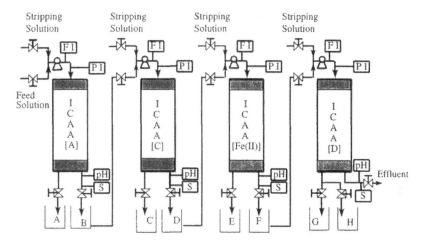

Figure 26 Integrated adsorption process for recovery of iron, copper, zinc, lead, and cadmium from Berkeley Pit waters. FI: flow indicator; PI: pressure indicator, pH: pH probe; S: sample port; ⊢⊠: solonoid valve. Tank A: concentrated Fe(III); B: bed 1 effluent; C: concentrated Cu(II); D: bed 2 effluent; E: concentrated Fe(II); F: bed 3 effluent; G: concentrated Zn/Cd; H: concentrated lead. (From Ref. 98.)

that the pH of the solution is increased to approximately 8–10, providing a nonacidic water for release to the environment.

ACKNOWLEDGMENTS

The financial support of the National Science Foundation through grant CTS-9805118 is gratefully acknowledged. We also acknowledge support of our previous research works by the U.S. EPA through grants R818639-01-1 and CR822727-0100.

REFERENCES

1. C Kantipuly, S Katragadda, A Chaw, HD Gesser. Talanta 37:491–517, 1990.
2. GR Pearson. J Am Chem Soc 85:3533–3539, 1963.

3. RA Khisamutdinov, NG Afzaletdinova, YI Murinov, EV Vasil'eva. Russian J Inorg Chem 42:1560–1566, 1997.
4. G Chessa, G Marangoni, B Pitteri, N Stevanato, A Vavasori. Reactive Polymers 14:143–150, 1991.
5. SJ Al-Bazi, A Chow. Talanta 31:815–836, 1984.
6. SD Alexandratos, AW Trochimczuk, DW Crick, EP Horwitz, RC Gatrone. Macromolecules 29:1021–1026, 1996.
7. FB Smith, WT Robinson, DG Jarvinen. Metal-ion separation and preconcentration—Progress and opportunities. ACS Symp Ser 716:294–330, 1999.
8. DG Jarvinen, FB Smith, WT Robinson. Metal separation technologies beyond 2000: Integrating novel chemistry with processing. Proceedings of a symposium sponsored by the Engineering Foundation Conference and National Science Foundation, Hawaii, 1999, pp 131–138.
9. G Cote, FM Chen, D Bauer. Solvent Extr Ion Exch 9:289–308, 1991.
10. T Braun, G Ghersini. Extraction Chromatography. Amsterdam: Elsevier, 1975.
11. KK Unger, N Becker, P Roumeliotis. J Chromatogr 125:115–127, 1976.
12. JR Jezorek, H Freiser. Anal Chem 51:366–373, 1979.
13. MA Marshall, HA Mottola. Anal Chem 55:2089–2093, 1983.
14. T Suzuki, K Tsunoda, H Akaiwa. Chem Lett 881–884, 1994.
15. GV Kudryavtsev, SZ Bernadyuk, GV Lisichkin. Russian Chem Rev 58:406–419, 1989.
16. RL Bruening, BJ Tarbet, KE Krakowiak, ML Bruening, RM Izatt, JS Bradshaw. Anal Chem 63:1014–1017, 1991.
17. RL Bruening, BJ Tarbet, RM Izatt, JS Bradshaw. Compositions and processes for removing and concentrating desired ions from solutions using sulfur and aralkyl nitrogen containing ligands bonded to inorganic supports. US patent no. 5,173,470, 1992.
18. LF Lindoy, P Eaglen. Ion complexation by silica-immobilized polyethyleneimines. US patent no. 5,190,660, 1993.
19. LL Tavlarides, NV Deorkar. Ceramic compositions with a pyrogallol moiety. US patent no. 5,624,881, 1997.
20. LL Tavlarides, NV Deorkar. Ceramic compositions with a phospho-acid moiety. US patent no. 5,612,175, 1997.
21. LL Tavlarides, NV Deorkar. Chemically active ceramic compositions with a thiol and/or amine moiety. US patent no. 5,616,533, 1997.
22. LL Tavlarides, NV Deorkar. Ceramic compositions with a hydroxyquinoline moiety. US patent no. 5,668,079, 1997.
23. X Feng, GE Fryxell, L-Q Wang, AY Kim, J Liu, KM Kemner. Science 276:923–926, 1997.

24. L Mercier, TJ Pinnavaia. Environ Sci Technol 32:2749–2754, 1998.
25. SD Alexandratos, DW Crick. Ind Eng Chem Res 35:635–644, 1996.
26. V Ravindran, MR Stevens, BN Badriyha, M Pirbazari. AIChE J 45:1135–1146, 1999.
27. M Rovira, L Hurtado, JL Cortina, J Arnaldos, AM Sastre. Solvent Extr Ion Exch 16:545–564, 1998.
28. NV Deorkar, LL Tavlarides. Emerging Separation Technologies for Metals II. Warrendale, PA: The Minerals, Metals & Materials Society, 1996, pp 107–118.
29. NV Deorkar, LL Tavlarides. Ind Eng Chem Res 36:399–406, 1997.
30. JS Kim, J Yi. J Chem Technol Biotechnol 74:544–550, 1999.
31. EP Plueddenmann. Silane Coupling Agents. 2nd ed. New York: Plenum Press, 1991.
32. GV Kudryavtsev, GV Lisichkin, VM Ivanov. Zhurnal Analiticheskoi Khimii 38:22–32, 1983.
33. JS Bradshaw, RL Bruening, KE Krakowiak, BJ Tarbet, ML Bruening, RM Izatt, JJ Christensen. J Chem Soc Chem Commun 812–814, 1988.
34. VI Fadeeva, TI Tikhomirova, IB Yuferova, GV Kudryavtsev. Anal Chim Acta 219:201–212, 1989.
35. TI Tikhomirova, MV Luk'yanova, VI Fadeeva, GV Kudryavtsev, OA Shpigun. Zhurnal Analiticheskoi Khimii 48:73–77, 1993.
36. PD Verweij, S Sital, MJ Haanepen, WL Driessen, J Reedijk. Eur Polym J 29:1603–1614, 1993.
37. JGP Espinola, JMPd Freitas, SFD Oliveira, C Airoldi. Colloids Surf A87:33–38, 1994.
38. C Airoldi, EFC Alcantara. J Chem Thermodyn 27:623–632, 1995.
39. R Chiarizia, EP Horwitz, KA D'Arcy, SD Alexandratos, AW Trochimczuk. Solvent Extr Ion Exch 14:1077–1100, 1996.
40. HJ Hoorn, Pd Joode, WL Driessen, J Reedijk. Recl Trav Chim Pays-Bas 115:191–197, 1996.
41. P Lessi, NLD Filho, JC Moreira JTS Campos. Anal Chim Acta 327:183–190, 1996.
42. AR Cestari, C Airoldi. J Colloid Interface Sci 195:338–342, 1997.
43. NLD Filho. J Colloid Interface Sci 206:131–137, 1998.
44. NLD Filho. Colloids Surf A144:219–227, 1998.
45. NLD Filho, Y Gushiken, DW Franco, MS Schultz, LCG Vasconcellos. Colloids Surf 141:181–187, 1998.
46. EM Soliman. Anal Lett 31:299–311, 1998.
47. NLD Filho. Colloid J 61:212–218, 1999.
48. NLD Filho. Polyhedron 18:2241–2247, 1999.

49. S Mattigod, GE Fryxell, X Feng, J Liu. Metal Separation technologies beyond 2000: Integrating novel chemistry with processing. Proceedings of a symposium sponsored by the Engineering Foundation Conference and National Science Foundation, Hawaii, June 1999, pp 71–79.
50. EM Rabinovich. J Mater Sci 20:4259–4297, 1985.
51. J Boilot, F Chaput, J Galaup, A Veret-Lemarinier, D Riehl, Y Levy. AIChE J 43:2820–2826, 1997.
52. S Parbakar, RA Assink. J Non-Cryst Solids 211:39–48, 1997.
53. BMD Witte, D Commers, JB Uytterhoeven. J Non-Cryst. Solids 202:35–41, 1996.
54. BV Zhmud, J Sonnefeld. J Non-Cryst. Solids 195:16–27, 1996.
55. NV Deorkar, LL Tavlarides. Hydrometallurgy 46:121–135, 1997.
56. P Zwietering. The Structure and Properties of Porous Solids. London: Butterworths, 1958.
57. FAL Dullien. Porous Media. New York: Academic Press, 1979.
58. CJ Brinker, KD Keeper, DW Schaefer, CS Ashley. J Non-Cryst. Solids 148:47–64, 1982.
59. KA Larson, JM Wiencek. Ind Eng Chem Res 31:2714–2722, 1992.
60. C-K Lee, LL Tavlarides. Ind Eng Chem Fundam 25:96–102, 1986.
61. K Sanjiv, V Rakesh, B Venkataramani, VS Raju, S Gangadharan. Solvent Extr Ion Exch 13:1097–1121, 1995.
62. JS Lee, NV Deorkar, LL Tavlarides. Ind Eng Chem Res 37:2812–2820, 1998.
63. J Ji, WC Cooper. Electrochimica Acta 41:1549–1560, 1996.
64. JW Ball, ND Kirk. J Chem Eng Data 43:895–918, 1998.
65. BE Reed, MR Matsumoto. Sep Sci Technol 28:2179–2195, 1993.
66. R-S Juang, R-H Lo. J Chem Tech Biotechnol 58:261–269, 1993.
67. RW Freeman, LL Tavlarides. J Inorg Nucl Chem 43:2467–2469, 1981.
68. E Hogfeldt. Stability Constants of Metal-Ion Complexes: Part A: Inorganic Ligands. 1st ed. New York: Pergamon Press, 1982.
69. WD Schecher, DC McAvoy. MINEQL+: A Chemical Equilibrium Program for Personal Computers, The Procter and Gamble Company, 1991.
70. CK Lee, LL Tavlarides. Polyhedron 4:47–51, 1985.
71. AG Sharpe. The Chemistry of Cyano Complexes of Transition Metals. London: Academic Press, 1976.
72. DC Grahame. Chem Rev 41:441–501, 1947.
73. CP Huang, MH Wu. Wat Res 11:673–679, 1977.
74. JA Davis, JO Leckie. J Colloid Interface Sci 67:90–107, 1978.
75. JA Davis, JO Leckie. Journal of Colloid and Interface Science 74:32–43, 1980.

76. H Tamura, E Matijevic, L Meites. J Colloid Interface Sci 92:303–314, 1983.
77. S Yiacoumi, C Tien. Kinetics of Metal Ion Adsorption from Aqueous Solutions: Models, Algorithms, and Applications. Norwell, MA: Kluwer Academic Publishers, 1995.
78. S Akita, H Takeuchi. J Chem Eng Jpn 23:439–443, 1990.
79. T Kataoka, H Yoshida. Chem Eng J 37:107–114, 1988.
80. C Tien. Adsorption Calculations and Modeling. Boston: Butterworth-Heinemann, 1994.
81. DD Do. Adsorption Analysis: Equilibria and Kinetics. London: Imperial College Press, 1998.
82. AB Mersmann, SE Scholl. Fundamentals of Adsorption. New York: United Engineering Trustees, 1991.
83. S Yiacoumi, C Tien. J Colloid Interface Sci 175:347–357, 1995.
84. BM v Vliet, WJ Weber Jr., H Hozumi. Wat Res 14:1719–1728, 1980.
85. BT Kim, HK Lee, H Moon, KJ Lee. Sep Sci Technol 30:3165–3182, 1995.
86. A Fernandez, C Suarez, M Diaz. J Chem Tech Biotechnol 58:255–260, 1993.
87. N Swami, DB Dreisinger. Solvent Extr Ion Exch 13:1037–1062, 1995.
88. A Warshawsky, MMB Fieberg, P Mihalik, TG Murphy, YB Ras. Sep Purif Methods 9:209–265, 1980.
89. A Sugii, N Ogawa, H Hashizume. Talanta 27:627–631, 1980.
90. SD Alexandratos, R Beauvais, JR Duke, BS Jorgensen. J Appl Polym Sci 68:1911–1916, 1998.
91. S Yiacoumi, C Tien. J Colloid Interface Sci 175:333–346, 1995.
92. H Lee, J Kim, J Yi. Chem Eng Commun (accepted, 2000).
93. JB Rosen. J Chem Phys 20:387–393, 1952.
94. JJ Perona, AC Coroneos, TE Kent, SA Richardson. Sep Sci Technol 30:1259–1268, 1995.
95. LK Fillippov. Chem Eng Sci 51:4013–4024, 1996.
96. GD Smith. Numerical Solution of Partial Differential Equations: Finite Difference Methods. 3rd ed. Oxford, England: Oxford University Press, 1985.
97. J Villadsen, ML Michelsen. Solution of Differential Equation Methods by Polynomial Approximation. Englewood Cliffs, NJ: Prentice Hall, 1978.
98. NV Deorkar, LL Tavlarides. Environ Prog 17:120–125, 1998.
99. Resource Recovery Project Technology Demonstrations—Series IV, No. A60488. Butte, MT: MSE Technology Applications, March 1, 1996.

6

Environmental Separation Through Polymeric Ligand Exchange

Arup K. SenGupta

Lehigh University, Bethlehem, Pennsylvania

I. INTRODUCTION

It is well recognized that the sorption affinity toward target solutes can be greatly enhanced by modifying and tailoring the interfacial chemistry of the sorbent. To this end, Helfferich [1,2] was the first to conceive the use of Cu(II)- or Ni(II)-loaded weak-acid cation exchange resins for ligands sorption through Lewis acid–base interaction. In such ligand-exchange processes, the water molecules (weak ligands) present at the coordination spheres of immobilized Cu(II) and Ni(II) in the cation exchange resins are replaced by relatively strong ligands, such as ammonia or ethylenediamine. The following provides a typical ligand-exchange reaction with ammonia, where M represents a divalent metal ion like Ni(II) or Cu(II) with strong Lewis acid properties and the overbar denotes the exchanger phase:

$$\overline{(RCOO^-)_2 M^{2+}(H_2O)_n} + nNH_3$$
$$\rightleftarrows \overline{(RCOO^-)_2 M^{2+}(NH_3)_n} + nH_2O \quad (1)$$

229

In addition to providing a quantitative approach toward determining ligand-exchange capacity of metal-loaded cation exchangers, Helfferich [2] also unveiled a striking similarity between ion-exchange and ligand-exchange processes. In heterovalent ion exchange, it is well known that with an increase in electrolyte concentration, the affinity (or binary separation factor) of the counterion with lower valence increases over the counterion with higher valence. In the realm of ion exchange, this phenomenon is popularly known as *electroselectivity reversal.* In a similar vein, Helfferich showed that the affinity of a monodentate ligand (ammonia) toward the exchanger in ligand-exchange processes is enhanced over a bidentate ligand (1,3-diaminopropanol) as the total ligand concentration is increased. This observation provided the basis for efficient regeneration of the bed at the end of the ligand-exchange process. During the last thirty years, much work has been done in applying the concept of ligand exchange in the areas of analytical chemistry, separation technology, and pollution control processes with varied amount of success [3–8].

The investigations in ligand exchange have, however, been confined primarily to nonionized (uncharged) ligands, namely, various amines and ammonia derivatives. In reality, most of the environmentally significant inorganic and organic ligands are anionic, such as cyanide, selenite, sulfide, acetate, oxalate, phthalate, phenolate, and naturally occurring humates and fulvates. Ligand-exchange processes using polymeric cation exchangers as metal hosts, as depicted in Eq. (1), are unable to sorb any anionic ligand, i.e.,

$$\overline{(RCOO^-)_2 M^{2+}(H_2O)_n} + \begin{cases} CN^- \\ SeO_3^{2-} \\ HS^- \\ CH_3COO^- \end{cases} \rightarrow \text{No reaction} \qquad (2)$$

The metal-loaded weak-acid cation exchange resins are electrically neutral and do not have any anion-exchange capacity, and, also, the negatively charged fixed co-ions (carboxylates in this case) of the polymer will not allow uptake of any anions in accordance with the Donnan co-ion exclusion principle. Thus, in spite of being strong ligands, the anions in Eq. (2) cannot displace water molecules (much weaker ligands) from the coordination spheres of the metal ions (Lewis acids). It is recognized that in order to sorb anionic ligands selectively, the polymeric substrate upon metal loading must possess fixed positive charges, i.e., it should act as an anion exchanger. Obviously, such functional polymers should have high preference toward metal ions so that the metals do not bleed or bleed only negligibly during the ligand-exchange process.

II. CHARACTERIZATION OF THE POLYMERIC LIGAND EXCHANGER AND GOALS

Conceptually, transition metal cations, say copper(II), if held firmly onto a solid phase at high concentrations, may act as anion-exchange sites with relatively high affinities toward aqueous phase anions with strong ligand characteristics. Figure 1 shows the major constituents of a polymeric ligand exchanger (PLE): first, a crosslinked polymeric template, like other anion exchangers; second, a chelating functional group covalently attached to that template; and, third, a Lewis acid–type metal cation strongly coor-

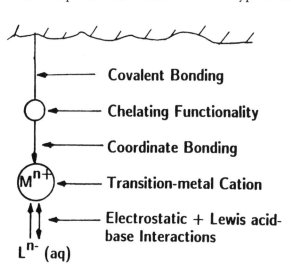

Figure 1 Various constituents of a conceptualized polymeric ligand exchanger for selective sorption of an aqueous phase ligand, L^{n-}. (From Ref. 24. Courtesy of the American Chemical Society.)

dinated to the chelating functional group in a manner that its positive charges are not neutralized. The resulting material is essentially polymer-anchored metal ions with fixed positive charges. Thus, from a generic viewpoint, the polymeric ligand-exchange process can be viewed as the formation of a ternary complex in the polymer phase, as shown here:

$$\overline{\overline{\overline{RL}}} \longrightarrow \overline{\overline{M^{n+}}} \; \text{----} \; \overline{L^{n-}}$$

$$(3)$$

where RL is the electrically neutral polymer-phase ligand, M^{n+} is the immobilized transition metal cation, and L^{n-} is the target anionic ligand. The overbars denote the polymer phase: one overbar indicates exchangeable anions participating in the ligand-exchange reactions, two overbars indicates the metal ions (Lewis acids) immobilized onto the polymeric substrate, and three overbars represents the covalently attached functional groups (Lewis bases) with no fixed charges. Coordination requirements of the metal ion (shown by arrows) are satisfied by both RL and L^{n-}, and at the same time electrostatic interactions (or ion pair formations) are operative between M^{n+} and L^{n-} (shown by dashed lines).

Specialty chelating polymers with nitrogen donor atoms satisfy the requisite properties to serve as the anchors of the transition metal cations. Matejka and Weber [9] used copper(II)- and chromium(III)-loaded chelating polymers containing oligoethyleneamine functional groups for removal of anionic ligands, namely, citrate, EDTA, and nitrilotriacetate (NTA). Competing effects of chloride and sulfate anions on ligand uptake were quite significant, and magnesium sulfate was used as the regenerant for desorption of ligand anions. Chanda et al. [10] studied sorption of arsenates and other ligands onto Fe(III)-loaded chelating polymers with nitrogen donor atoms. Iron is a nontoxic, innocuous metal and hence, its bleeding from the polymer phase does not pose any environmental hazard. However, Fe(III) is a hard Lewis acid with poor affinity toward functional groups containing nitrogen donor atoms. The arsenate removal capacity in the presence of competing sulfate and chloride was rather low.

The primary subject of this chapter pertains to the use of a specific class of commercially available chelating exchangers containing only nitrogen donor atoms (bis-picolylamine functional groups) for ligand exchange. These chelating exchangers are macroporous with polystyrene matrix and divinylbenzene crosslinking, and are available in spherical bead forms from Dow Chemical Co. (Midland, MI). These exchangers have extremely high affinities toward Cu(II) ions and are ideally suited for anchoring Cu(II) [11–15]. Such tailored copper-loaded chelating exchangers provide an excellent sorbent for anionic ligands and are henceforth referred to as polymeric ligand exchangers.

Table 1 includes two commercially available chelating exchangers with functional groups containing only nitrogen donor atoms without any fixed charges. The exchanger XFS 4195 contains three nitrogen donor atoms per repeating monomer and is referred to as DOW 3N in this chapter, while XFS 43084 containing two nitrogen atoms per repeating unit is designated as DOW 2N. In order to develop an insight into the underlying sorption mechanism, two strong-base anion exchangers with quaternary ammonium functional groups and one popular chelating exchanger with iminodiacetate functional group are also included in the study. Table 1 includes their salient properties and the names of the manufacturers.

DOW 3N and DOW 2N are converted into copper-loaded forms by passing 500 mg/L Cu(II) solution through them in separate columns until saturation. Cupric sulfate and cupric chloride solutions are used. The resins are subsequently rinsed and air dried for 72 h. For typographical convenience, these two polymeric ligand exchangers will also be referred to as DOW 2N-Cu and DOW 3N-Cu.

Many environmental and process applications require selective separation of trace concentrations (µg/L to mg/L) of anionic ligands (namely, oxalate, phosphate, phthalate, ethylenediaminetetraacetate (EDTA), arsenates, nitrilotriacetate (NTA), selenites, and cyanides) onto a suitable sorbent from the background of high concentrations of competing anions (namely, chloride and sulfate). Ligand characteristics and other related properties of some of these solutes are provided in Table 2. Phosphate at trace concentration is responsible for eutrophication (algal bloom); selenium and arsenic are included in US EPA's list of priority pollutants; and oxalate is a model compound for various hydrophilic organic anions. Several previous studies [16–22] have confirmed the need to identify and characterize appropriate sorbents which would exhibit high affinities toward selenite, phosphate, and low molecular weight organic anions in the presence of other electrolytes. The specific goals of this chapter are (1) to present convincing experimental evidence in support of PLE's high affinity toward target ligands; (2) elucidate the underlying sorption mechanism, and (3) provide regeneration methodologies following exhaustion.

Table 1 Background Information on Ion Exchangers

Composition of the functional group	Characteristic	Matrix, porosity	Manufacturer and trade name
Ⓡ—⟨N⟩—CH₂–N:–CH₂—⟨N⟩	High metal ion affinity	Polystyrene, macroporous	DOW Chemical:[a] DOW 3N or XFS 4195
Ⓡ—⟨N⟩—CH₂–N:–CH₂–CHOH–CH₃	High metal ion affinity	Polystyrene, macroporous	DOW Chemical: DOW 2N or XFS 43084
Ⓡ—⟨N⟩—CH₃–N⁺–CH₃ / CH₃	Strong-base anion exchanger	Polystyrene, macroporous	Rohm and Haas, Co.:[b] IRA-900
		Polyacrylic, macroporous	Rohm and Haas, Co.: IRA-958
Ⓡ—N:⟨CH₂COO:⁻ / CH₂COO:⁻	High metal ion affinity	Polystyrene, macroporous	Rohm and Haas, Co.: IRC 718

[a] Dow Chemical Co., Midland, MI.
[b] Rohm & Haas, Co.,
Note: Circled R denotes the repeating unit of the polymer matrix.

Table 2 Properties of Target and Competing Anions

Anion	Ligand characteristic	Parent acid	pK_a values[a]
SeO_3^{2-} (selenite)	Bidentate ligand with two oxygen donor atoms	H_2SeO_3	$pK_1 = 2.6$ $pK_2 = 8.3$
HPO_4^{2-} (phosphate)	Bidentate ligand with two oxygen donor atoms	H_3PO_4	$pK_1 = 2.1$ $pK_2 = 7.2$ $PK_3 = 12.4$
$HAsO_4^{2-}$ (arsenate)	Bidentate ligand with two oxygen donor atoms	H_3AsO_4	$pK_1 = 2.2$ $pK_2 = 7.0$ $pK_3 = 11.5$
$C_2O_4^{2-}$ (oxalate)	Bidentate ligand with two oxygen donor atoms	$H_2C_2O_4$	$pK_1 = 1.3$ $pK_2 = 4.3$
$C_6H_4(COO)_2^{2-}$ (ortho-phthalate)	Bidentate ligand with two oxygen donor atoms	$C_6H_4(COOH)_2$	$pK_1 = 2.9$ $pK_2 = 5.4$
SO_4^{2-} (sulfate)	Very poor ligand with ability to form ion pairs with metal cations	H_2SO_4	$pK_1 = $ negative $pK_2 = 1.9$
Cl^- (chloride)	Weak monodentate ligand	HCl	$pK = $ negative

[a] Values from JA Dean, ed. Lange's Handbook of Chemistry. New York: McGraw-Hill, 1979.

III. RESULTS, PERFORMANCE BENCHMARKING, AND MODELING

A. Isotherms and Effects of Competing Ions

Municipal and industrial wastewater always contains sulfate and chloride anions which will compete with target phosphate for the sorption sites. Of these anions, sulfate possesses higher ionic charges (i.e., divalent) and will offer greater competition through enhanced electrostatic interaction. To assess the competing effects of sulfate on the PLE's phosphate uptake, isotherm tests using the minicolumn technique were conducted at two different background sulfate concentrations, namely, 200 and 400 mg/L, with other conditions remaining identical. Figure 2 shows the phosphate

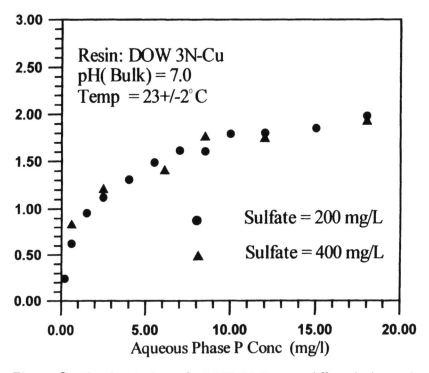

Figure 2 Phosphate isotherms for DOW 3N-Cu at two different background concentrations (200 and 400 mg/L) of competing sulfate ions. (From Ref. 23. Courtesy of IWA.)

uptake for these two separate isotherm tests. Note that doubling the concentration of the competing sulfate ion had an insignificant effect on the phosphate removal capacity of the PLE.

Phosphate–sulfate separation factors, as computed from the experimental data, are provided in Fig. 3. Also superimposed in Fig. 3 are the experimentally determined separation factors of other anion exchangers studied previously [21–23]. A separation factor is a measure of relative selectivity between two competing solutes and is equal to the ratio of their distribution coefficients between the exchanger phase and the aqueous phase. Thus, the phosphate/sulfate separation factor ($\alpha_{P/S}$) can be expressed as

$$\alpha_{P/S} = \frac{q_P}{C_P} \frac{C_S}{q_S} \tag{4}$$

where q and C represent the concentrations (molar or equivalent) in the polymer phase and aqueous phase, and subscripts P and S denote phos-

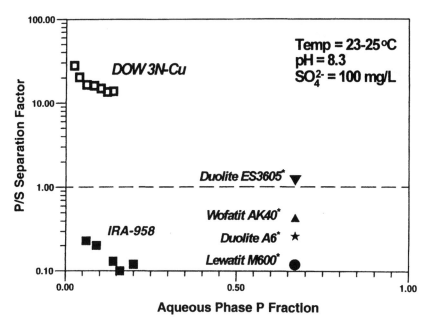

Figure 3 Comparison of phosphate/sulfate (P/S) separation factors for various sorbents. (From Ref. 23. Courtesy of IWA.)

phate and sulfate, respectively. Although both phosphate and sulfate exist as divalent anions at pH 8.3, the average separation factor, $\alpha_{P/S}$, with PLE is well over an order of magnitude greater compared to IRA-958 and other sorbents studied to date.

Figure 4 shows a comparison of Se(IV)/sulfate isotherm data for DOW 2N-Cu and IRC 718-Cu at a pH of 7.0, where $HSeO_3^-$ is the predominant Se(IV) species. The following may be noted:

1. The average selenite/sulfate separation factor for DOW 2N-Cu is around 10, indicating higher selectivity of monovalent $HSeO_3^-$ over SO_4^{2-}. Such an observation of higher selectivity of monovalent $HSeO_3^-$ over divalent SO_4^{2-} in a fairly dilute solution (ionic strength of 0.0078 M) is quite uncommon in the realm of ion exchange.

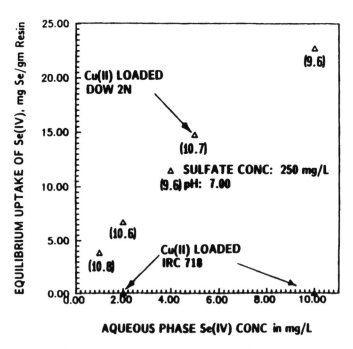

Figure 4 Binary selenite/sulfate isotherms for DOW 2N-Cu and IRC 718-Cu under identical experimental conditions. Numbers in parentheses indicate selenite/sulfate separation factors for DOW 2N-Cu at pH 7.0. (From Ref. 25. Courtesy of the American Society of Civil Engineers.)

2. Unlike DOW 2N-Cu, the other copper-loaded chelating exchanger, IRC 718-Cu, showed negligible selenite uptake under otherwise identical conditions. This latter observation with IRC-718 Cu (i.e., no selenite uptake) suggests that the presence of copper(II) in the polymer phase alone does not ensure high affinities toward Se(IV) oxyanions. Both DOW 2N-Cu and DOW 3N-Cu also exhibit high affinities for oxalate, sulfide, and arsenate [24,25].

B. Fixed-Bed Column Runs

1. Oxalate Removal

Figure 5 shows effluent histories of oxalate during separate column runs under identical conditions with four different sorbents: an activated carbon (Filtrasorb 300, Calgon Corp.), a strong-base polymeric anion ex-

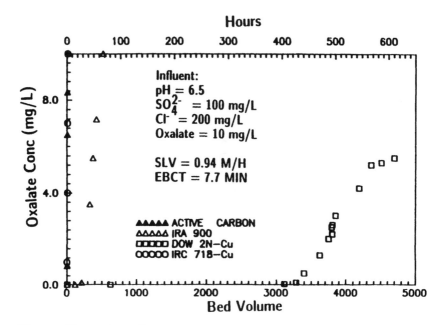

Figure 5 Oxalate effluent histories during column runs with four different sorbents under otherwise identical conditions: SLV, superficial liquid-phase velocity, EBCT, empty bed contact time. (From Ref. 24. Courtesy of the American Chemical Society.)

changer (IRA-900, Rohm and Haas, Co.), IRC 718-Cu, and the polymeric ligand exchanger, DOW 2N-Cu. Compared to other sorbents, oxalate breakthrough for DOW 2N-Cu occurred much later and after 3000 bed volumes, although competing sulfate and chloride were present in the influent at much greater concentrations. In contrast, activated carbon and IRC 718-Cu did not offer practically any oxalate removal capacity, while the strong-base anion exchanger (IRA-900) was completely exhausted in less than 250 bed volumes.

In order to avoid any possible bleeding of copper from the PLE bed into the exit of the column, a small amount (about 10% of the total bed height) of a virgin chelating ion exchanger (IRC-718, Rohm and Haas Co.) in sodium form was kept at the bottom of the column. Figure 6 provides a general sketch of the fixed-bed column used in the study. IRC-718 is a chelating cation exchanger with iminodiacetate functional group and polystyrene matrix; its salient properties are provided in Table 1.

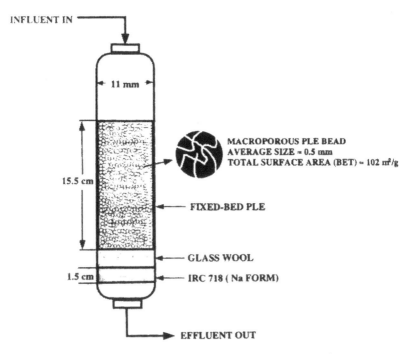

Figure 6 Details of the laboratory column used in the study.

2. Phosphate Removal

Selective phosphate removal from secondary wastewater is gaining importance in many industrialized nations [21–23,26–32]. The long-term
viability of the proposed process using PLE is, however, influenced by
the nature and intensity of interactions between dissolved organic matters
(DOMs) in the secondary wastewater and the sorption sites of the PLE.
In a typical secondary wastewater after biological treatment, the DOMs
are derived primarily from the residuals of protein, carbohydrates, fats,
and soluble microbial products [33–35]. Understandably, the hydrophobic, electrostatic, and ligand properties of these individual DOM molecules will influence their abilities to compete for sorption sites in the PLE.
In order to monitor the direct impact of DOMs on phosphate removal
by DOW 3N-Cu, a lengthy column run was carried out using secondary
wastewater obtained from the Bethlehem Wastewater Treatment Plant
(BWTP).

Figure 7 provides the effluent histories of phosphate and dissolved
organic carbon (DOC) during the column run. The composition of the

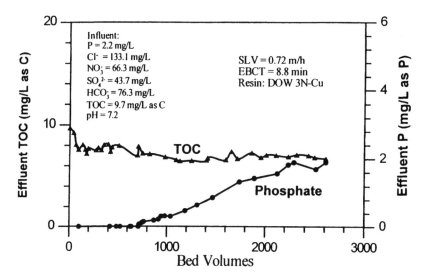

Figure 7 Total organic carbon (TOC) and phosphate breakthrough curves
during the treatment of secondary wastewater from Bethlehem Wastewater Treatment Plant using DOW 3N-Cu. (From Ref. 23. Courtesy of IWA.)

secondary wastewater is also provided in Fig. 7. Phosphate concentration was 2.2 mg/L as P, while that of DOC was 9.7 mg/L. Note that while phosphate breakthrough took place after about 800 bed volumes, DOC was poorly adsorbed and broke through almost immediately. Also, no chromatographic elution was observed for phosphate breakthrough, i.e., the phosphate concentration at the exit of the column after breakthrough never exceeded its influent concentration. This observation confirms that DOMs and other anions commonly present in secondary wastewater have a lower affinity toward the PLE compared to phosphate. Earlier pilot plant studies with strong-base anion exchangers, however, revealed conclusively that sulfate and certain components of DOMs are preferred over phosphate, thus, demanding frequent regenerations [27,28,32]. PLE thus offers a unique opportunity to selectively remove phosphate from secondary wastewater using a simple-to-operate fixed-bed column.

C. Column Behavior

During one of the fixed-bed column runs with PLE (DOW 3N-Cu) for phosphate removal from a representative secondary wastewater, the flow of influent to the column was deliberately stopped after about 1240 bed volumes. At this point, the effluent phosphate concentration was approximately 1.1 mg/L as phosphorus (P), or 25% of the phosphate present in the influent. Six hours later, the original flow conditions were restored, and samples were collected and analyzed for phosphate. Figure 8 shows that the phosphate concentration at the exit, after 6-h interruption, dropped to less than 0.5 mg/L as P. After the re-start of the column, the phosphate concentration gradually rose to 1.2 mg/L as P. However, the overall phosphate removal capacity of the fixed bed remained unchanged from the 6-h interruption.

For an intraparticle diffusion–controlled sorption process, the concentration gradient within an exchanger particle serves as the driving force and governs the overall rate. With the progress of the column run, this concentration gradient attenuates. The interruption allows the sorbed phosphate to evenly spread out inside the exchanger bed. As a result, the concentration gradient, and thus the uptake rate immediately after the column re-start, is greater than the uptake rate prior to the interruption. In other words, a faster uptake immediately after the interruption provides

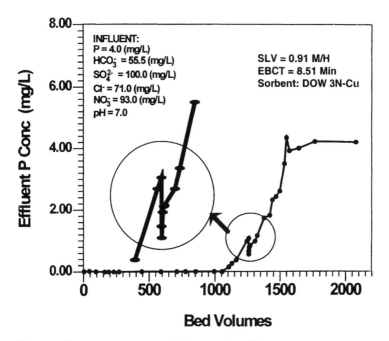

Figure 8 Demonstration of a drop in the effluent concentration of phosphate following a 6-h column run interruption.

evidence in support of intraparticle diffusion being the rate-limiting step. Figure 9 depicts the changes in concentration gradients during various stages of interruption test. The dimensionless Reynolds number, Re_p, under the experimental conditions of the column run in Fig. 8 was very low, less than 0.2. Thus, at higher liquid phase velocity or higher Re_p values, the intraparticle diffusion will continue to be the rate-limiting step.

As already shown, the sorption isotherms of the target ligands with PLE are highly favorable, i.e., convex upward. Also, intraparticle diffusion is the rate-limiting step and the effective intraparticle diffusivity is relatively low due to the target solute's high affinity toward PLE. Thus, the breakthrough of the target ligand follows a constant pattern behavior, i.e., the length of the mass transfer zone does not depend on the length of the column.

Figure 10 shows phosphate breakthrough curves for column runs with three different bed heights but under otherwise identical conditions.

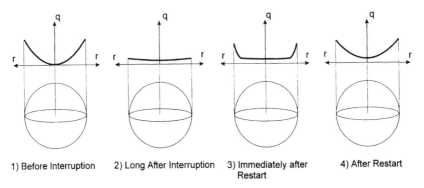

1) Before Interruption   2) Long After Interruption   3) Immediately after Restart   4) After Restart

Figure 9 Concentration gradients within a single exchanger spherical bead during the various stages of the interruption test.

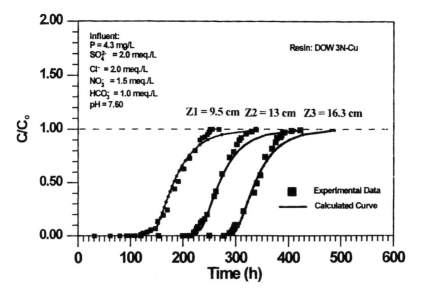

Figure 10 Phosphate breakthrough curves at three different bed depths, validating the constant breakthrough profile. (The solid lines are model predicated curves. Intraparticle diffusivites are concentration dependent. At high phosphate concentrations, the breakthrough curves depart from model predictions.)

The resemblance of breakthrough patterns at varying bed heights validates the constant pattern breakthrough profile. Relatively simple mathematical models based on linear concentration gradients can well predict constant pattern profiles. The solid lines in Fig. 10 represent the model predictions. Detailed mathematical deductions and associated computations are available elsewhere [36]. Note that the effective intraparticle diffusivity is concentration dependent; it is greater at higher concentration of phosphate. The constant pattern breakthrough model used does not take this effect into account. That is why at near saturation the phosphate breakthrough curve is sharper, i.e., the model underpredicts the mass transfer–limited kinetics.

D. Underlying Mechanism

Column run results clearly demonstrate that the polymeric ligand exchanger offered much higher phosphate and oxalate removal capacities compared to strong-base anion exchangers (IRA-958 and IRA-900) and IRC 718-Cu. Figure 11 provides a schematic presentation of the underlying mechanisms which govern the binding of a bidentate anion (e.g., phosphate) onto these three sorbents. Considering the fact that the copper(II) ion has four primary coordination numbers, Fig. 11 illustrates the following:

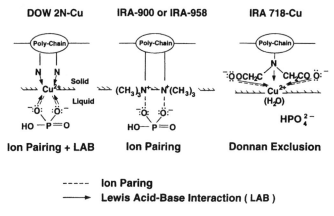

Figure 11   Schematic presentation of binding mechanisms for different sorbents.

1. At around neutral to slightly alkaline pH, two coordination numbers for Cu(II) are satisfied by the two nitrogen donor atoms in the polymer phase and, as a result, copper ions are held firmly onto these sites of the PLE. The remaining two coordination numbers and the two residual positive charges of the polymer phase copper(II) ion are satisfied simultaneously by the divalent phosphate anion (HPO_4^{2-}) with two oxygen donor atoms. Thus, coulombic interaction (i.e., ion pair formation for maintenance of electroneutrality) is accompanied by Lewis acid–base interaction where electron deficiencies in the coordination spheres of Cu^{2+} (Lewis acid) are satisfied by donor oxygen atoms of phosphate.

2. For IRA-900 or IRA-958, on the contrary, only columbic interaction (ion pair formation) is involved between the positively charged quaternary ammonium group (R_4N^+) and the anion (HPO_4^{2-}) because R_4N^+ does not have any electron-acceptor characteristic. As a result, the sorption affinity of HPO_4^{2-} toward IRA-958 or other strong-base anion exchangers cannot be enhanced by Lewis acid–base interaction.

3. For IRC 718-Cu, the three coordination numbers of Cu(II) are satisfied by the iminodiacetate functional group (two oxygen and one nitrogen donor atoms) and, also, its positive charges are neutralized within the polymer phase by the acetate groups. Thus, IRC 718-Cu does not have any available anion-exchange capacity to bind an anion onto it, and the fourth coordination number of Cu(II) is satisfied by a neutral water molecule. In fact, according to the Donnan exclusion principle, anions are rejected by IRC-718 because of its negatively charged (iminodiacetate) functional group. In spite of high Cu(II) affinity, IRC 718-Cu cannot, therefore, sorb anions regardless of their ligand characteristics. However, sorption of nonionized ligands like ammonia and ethylenediamine onto IRC 718-Cu is possible.

The foregoing analysis remains equally valid to explain high affinities of selenite, arsenate, oxalate, and other anionic ligands toward the PLE. It has also been recently demonstrated that Cr(VI) anions can be very selectively removed at above-neutral pH with the PLE [37].

E. Modeling Equilibrium Behavior

Figure 12 shows that in a phthalate–sulfate system, the total capacity, Q, of DOW 3N-Cu in milliequivalents per gram remains practically con-

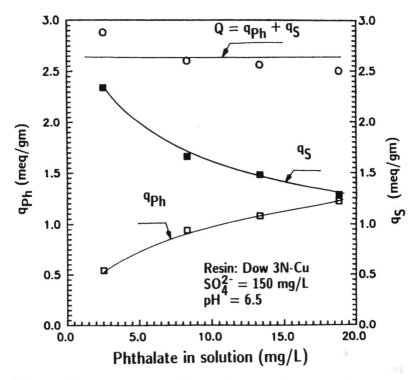

Figure 12 Demonstration of ion exchange stoichiometry with DOW 3N–Cu for phthalate–sulfate exchange. (From Ref. 24. Courtesy of the American Chemical Society.)

stant, i.e., the sorption/desorption process occurs on the equivalent exchange basis. Similar observations were also made for oxalate–sulfate and phosphate–sulfate exchange. Thus, the exchange reaction between sulfate and a divalent organic ligand (L^{2-}) with a PLE can be presented as follows:

$$\overline{PLE\text{-}Cu^{2+}(SO_4^{2-})} + L^{2-} \rightleftarrows \overline{(PLE\text{-}Cu^{2+})\,(L^{2-})} + SO_4^{2-} \qquad (5)$$

Assuming ideality in both aqueous and polymer phase, the equilibrium constant for the this reaction is given by

$$K = \frac{q_L}{q_S}\frac{C_S}{C_L} \qquad (6)$$

Subscripts L and S correspond to the ligand and sulfate, respectively. Considering all concentrations to be in equivalent units (i.e., milliequivalents per liter or milliequivalents per gram) and the fact that

$$q_L + q_S = Q \text{ (total capacity of PLE)} \tag{7}$$

and

$$C_S + C_L = C° \text{ (total equivalent aqueous phase concentration)} \tag{8}$$

Equation (6) takes the following form:

$$K = \frac{q_L}{Q - q_L} \frac{C° - C_L}{C_L} \tag{9}$$

First, let us consider the situation where $C°$ is constant but the ligand (L^{2-}) is a trace species compared to competing sulfate. Hence, $C_L \ll C°$ and $q_L \ll Q$. Therefore, $C° - C_L \simeq C°$ and $Q - q_L \simeq Q$. Rearrangement of Eq. (9) under these simplifying conditions gives

$$q_L = \frac{KQ}{C°} C_L \tag{10}$$

Since K, Q, and $C°$ are constant, q_L and C_L are linearly dependent on each other, i.e., it is the case of a linear isotherm. Now, for the situation where C_L and C_S are comparable and L^{2-} has much higher affinity compared to sulfate (i.e., high K value), rearrangement of Eq. (9) provides the following:

$$q_L = \frac{K C_L Q}{C° + (K - 1) C_L} \tag{11}$$

Since $K \gg 1$, $(K - 1)C_L \gg C°$, and $K - 1 \simeq K$, from Eq. (11),

$$q_L \simeq Q \tag{12}$$

In essence, when the target ligand is a trace species (i.e., C_L is very low), the sorption capacity, q_L, is linearly dependent on C_L. For much higher concentrations of C_L, however, q_L approaches the total capacity, Q. Characteristically, the behavior conforms to the Langmuir isotherm. Equation

11 can be further refined to an approximate Langmuir isotherm by expressing aqueous phase solute concentration in normalized equivalent fraction, which is

$$x_L = \frac{C_L}{C^\circ} \tag{13}$$

Equation (11) now becomes

$$q_L = \frac{QKx_L}{1 + (K - 1)x_L} \tag{14}$$

Considering K to be much greater than unity, Eq. (14), upon linearization, takes the following form:

$$\frac{1}{q_L} = \frac{1}{KQ}\frac{1}{x_L} + \frac{1}{Q} \tag{15}$$

The experimental oxalate isotherm data for DOW 3N-Cu, when fitted to Eq. (15) by plotting $1/q_L$, versus $1/x_L$, provide excellent agreement (Fig. 13).

As stated earlier, the polymeric ligand-exchange process may be envisioned as the formation of a ternary complex at the sorption sites of the chelating polymers with nitrogen donor atoms. Since Lewis acid–base interaction between the immobilized Cu(II) and target anionic ligand is the primary determinant of the sorption affinity, the linear free energy relationship (LFER) can be a useful tool in predicting the relative selectivity of PLE among different anions. For example, let us consider the following aqueous phase complexation reaction between Cu(II) and L^{2-}:

$$Cu^{2+} \text{ (aq)} + L^{2-} \text{ (aq)} = CuL^\circ \text{ (aq)} \tag{16}$$

Now the stability or formation constant of the above reaction is given by

$$K_f = \frac{CuL^\circ}{[Cu^{2+}]\,[L^{2-}]} \tag{17}$$

For a number of divalent anionic ligands, formation constants in the aqueous phase are available in the open literature. LFER predicts a strong correlation between the relative selectivity (say, with respect to sulfate)

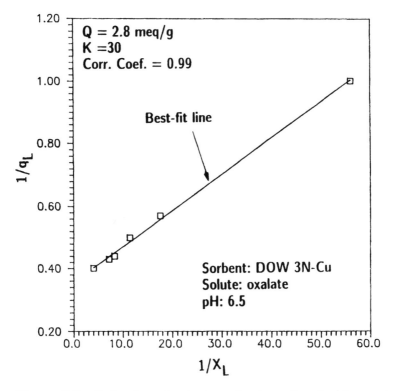

Figure 13 Fitting the oxalate isotherm data into a linearized Langmuir-type model.

and the aqueous phase formation constants for various anonic ligands [36]. Figure 14 plots the experimentally determined log $\alpha_{L/S}$ (logarithm of separation factor of the ligand with respect to sulfate) versus log K_f for a number of divalent ligands; a strong correlation can be noted between K_f and $\alpha_{L/S}$ in accordance with the predictions of LFER.

F. Regenerability

In order to be economically viable, a selective sorbent also needs to be amenable to efficient regeneration and reuse for multiple cycles. In reality, however, the efficiency of regeneration (or desorption) tends to diminish for highly selective sorbents. To strike a balance between selectivity and

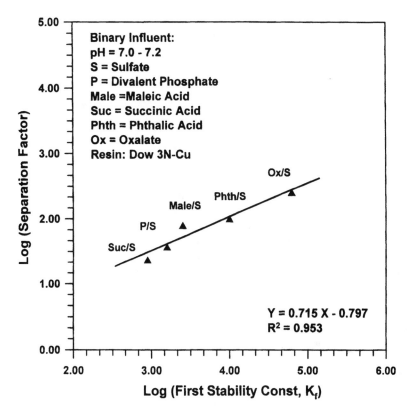

Figure 14 Linear free energy relationship (LFER) between ligands' separation factors with respect to sulfate and meta–ligand formation constants. (From Ref. 41. Courtesy of the American Chemical Society.)

regenerability, the intensity of solute–sorbent interaction has to lie within an envelope where ion exchange–type sorption is selective yet reversible and the sorption/desorption cycle is viable. Figure 15 helps quantify such a working regime for various types of interactions based on absolute free energy changes. Note that ligand exchange, although selective, falls within the reversible regime of the ion exchange process.

Following some experimental trial and error, 6% NaCl at pH 4.3 was found to be an efficient (and also inexpensive) regenerant for the exhausted PLE. Figure 16 shows the phosphate concentration profile during regeneration following a lengthy column run. Note that the phosphate desorption was very efficient and that approximately 96% of phosphate

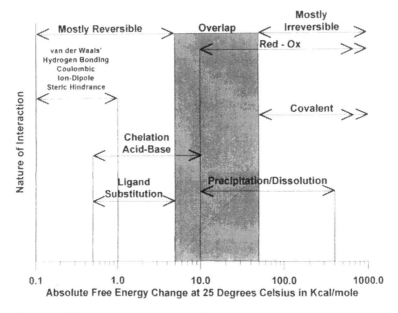

Figure 15 Range of reversibility for various interactions associated with ion exchange–type sorption processes.

was recovered from the bed in less than eight bed volumes. It is postulated that the following are the most predominant exchange reactions during the regeneration process:

$$\overline{R\text{-}Cu^{2+}\ (HPO_4^{2-})} + 2Cl^- + H^+ \rightleftarrows \overline{R\text{-}Cu^{2+}\ (Cl^-)_2} + H_2PO_4^-$$
(18)

$$\overline{R\text{-}Cu^{2+}\ (HPO_4^{2-})} + 2Cl^- \rightleftarrows \overline{R\text{-}Cu^{2+}\ (Cl^-)_2} + HPO_4^{2-} \qquad (19)$$

where R represents the repeating unit of the PLE with a bi-pyridyl functional group.

Both electroselectivity reversal at the high ionic strength of the regenerant (6% NaCl) and lower affinity of monovalent phosphate, $H_2PO_4^-$, at the prevailing pH are the underlying reasons for efficient phosphate regeneration. According to the selectivity reversal phenomenon, the selectivity of a monovalent counterion (chloride in this case) toward the

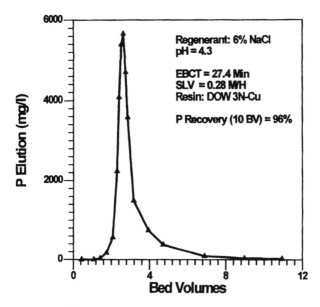

Figure 16 Phosphate elution profile during regeneration of the saturated DOW 3N-Cu using brine. (From Ref. 23. Courtesy of IWA.)

exchanger is enhanced compared to counterions of higher valences (HPO_4^{2-} or SO_4^{2-} in this case) at high aqueous phase electrolyte concentrations. The exit pH during the entire period (about 10 bed volumes) of the co-current regeneration remained between 6.0 and 7.0. Note that this pH is higher than that in the influent; this phenomenon is attributable to the residual elution of sorbed HCO_3^- and OH^-.

Chromate desorption using brine was also very efficient [37]. A two-step regeneration methodology using ammonia has been developed in which Cu(II), the ligand-exchange site, is first eluted irreversibly and then reloaded onto the chelating exchangers with nitrogen donor atoms. Although the regeneration process requires an additional weak-acid cation exchanger and involves two separate steps, the regeneration efficiency is very high. There is also no copper loss from the system. The details pertaining to individual steps of the regeneration process are provided by SenGupta and Zhu [38].

G. Copper Bleeding and Role of Copper Affinity

In order for the polymeric ligand-exchange process to be operationally viable, the bleeding of Cu(II) from the polymer phase during both service and regeneration cycles has to be minimal, if not absent altogether. During the lengthy column runs with the PLE and subsequent regenerations with brine, about 10% of virgin chelating exchanger (IRC-718) at the bottom of the bed was sufficient to eliminate any leakage of Cu(II) from the bed. Visually, it was clear that the virgin IRC-718 was far from being saturated with Cu(II) (note that upon copper loading, the chelating resin turns blue). Since regeneration conditions were far more extreme (6% NaCl), tests were carried out to quantitatively measure the relative amount of copper eluted from the bed during a typical regeneration cycle of the PLE. Ten bed volumes of spent 6% NaCl regenerant were passed through an excess of virgin IRC-718 to intercept the entire amount of copper bled from DOW 3N-Cu. Following acid regeneration of IRC-718 and mass balance on copper, it was estimated that the copper eluted from the PLE during regeneration amounted to 0.0056% of the total copper loaded onto DOW 3N-Cu. From an engineering design viewpoint, 10% virgin IRC-718 at the bottom of the column is quite sufficient to arrest copper bleeding from the PLE for over one hundred cycles of operation. Intermittently, after every hundred cycles, IRC-718 can be regenerated with dilute acid and the eluted copper can be reloaded onto the PLE. A previous study confirmed the high affinity of copper(II) toward DOW 3N even under highly acidic conditions [14,39].

Extremely low copper bleeding from the PLE during both exhaustion and regeneration cycles is a confirmation of the fact that commonly encountered cations, namely, Na^+ and Ca^{2+}, have negligible competing effects toward the parent chelating exchanger, DOW 3N. Sodium and calcium belong to Group IA and IIA in the periodic table, respectively. The electronic configurations of cations Na^+ and Ca^{2+} correspond to noble gas structures, namely, neon and argon. These cations (Na^+ and Ca^{2+}) are essentially hard spheres and can be sorbed only to negatively charged sites through coulombic interactions. However, the active chelating functional groups of DOW 3N have only nitrogen donor atoms with no negative charges, as shown in Table 1. This intrinsic property of DOW 3N (i.e., lack of negative surface charges) is the underlying reason for their lack of affinity toward hard cations like Ca^{2+}, Mg^{2+}, and Na^+. In contrast, the electronic configuration of copper(II) is $(Ar)3d^9$, which is far removed from a noble gas configuration. Compared to Group IA and IIA ions, copper(II) is essentially a soft cation with strong Lewis acid characteristic. Consequently, copper(II) undergoes strong Lewis acid–base interaction with the nitrogen donor atoms of DOW 3N, leading to high affinity even under acidic conditions [11,14]. Hard cations are unlikely to displace copper(II) from the PLE, and that is why any fugitive copper bleeding during lengthy column runs and regenerations was practically absent.

IV. CONCLUSION

One of the primary challenges in many environmental separation processes involves the selective removal of dissolved trace contaminants from a background of relatively high concentrations of competing but innocuous solutes [40]. Fixed-bed sorption processes are quite effective under such circumstances provided that the sorbents are selective toward the target contaminants. Also, the sorbents need to be amenable to efficient regeneration and reuse. It is recognized that solute–sorbent interactions may be altered and the sorption affinity enhanced by manipulating the chemistry at the sorption sites of the polymeric sorbents. This chapter discusses a new class of tailored sorbents referred to as polymeric ligand exchangers which exhibit very high affinities toward a host of environmentally significant anions. These new sorbents were also found very durable with high regeneration efficiency. The major attributes of the polymeric ligand exchangers can be summarized as follows:

Chelating polymers with nitrogen donor atoms, once loaded with copper(II), exhibit high affinities toward many environmentally significant anionic ligands. Immobilized Cu(II) ions are firmly retained onto the polymer phase and practically unaffected by the presence of other commonly encountered cations, namely, Na^+, Ca^{2+}, and Mg^{2+}.

The polymeric ligand exchangers are very effective for selective separation of trace solutes (phosphate, selenite, oxalate, chromate arsenate, and others) from water and wastewater. The underlying mechanism contributing to high sorption affinity comprises two major interactions:

first, an electrostatic or coulombic interaction resulting from the positively charged, immobilized Cu^{2+} and the solute anion; second, a strong Lewis acid–base interaction between Cu(II) and the solute's donor atoms.

The ligand exchange process conforms to ion exchange strichiometry, i.e., the sorption of a counterion is coupled with the desorption of an equivalent amount of other counterions. The sorption of a trace ligand onto PLE follows the characteristics of a Langmuir isotherm. Kinetically, intraparticle diffusion is the rate-limiting step. During fixed-bed column runs, the effluent histories of trace ligands follow "constant pattern breakthrough" behaviors and can be modeled using algorithms already existing in the open literature.

The new exchanger is extremely durable, chemically stable, and mechanically strong. A single-step brine regeneration at pH -4.5 was found effective for desorption of phosphate and other ligands. A two-step regeneration procedure, although operationally more complex, was very efficient and able to recover oxalate almost stoichiometrically in less than ten bed volumes.

ACKNOWLEDGMENTS

This study could not have been possible without the dedicated work of my former graduate students, Yuewei Zhu, Anu Ramana, and Dongye Zhao. Of them, Dr. Zhao's contribution was the most significant. I also acknowledge the financial support received from the United States Environmental Protection Agency toward this project.

REFERENCES

1. F Helfferich. Nature 189:1001, 1961.
2. F Helfferich. J Am Chem Soc 84:3237, 1962.
3. RA Dobbs, S Ushida, LM Smith, JM Cohen. AIChE Symp Ser 75:157, 1975.
4. CM de Hernandez, HF Walton. Anal Chem 46:890, 1972.
5. VA Davankov, AV Semechkin. J Chromatogr 141:313, 1977.
6. CY Liu, et al. Fres J Anal Chem 339:877, 1991.
7. HF Walton. Ind Eng Chem Res 34:2553, 1995.
8. FR Groves, T White. AIChE J 30:3,494–496, 1984.

9. Z Matejka R, Weber. React Polym 13:299, 1990.
10. M Chanda, KF O'Driscoll. GL Rempel. React Polym 9:277, 1988.
11. AK SenGupta, Y Zhu, D Hauze. Environ Sci Technol, 25:481, 1991.
12. RR Grinstead. Hydrometallurgy 12:387–400, 1984.
13. RR Grinstead. J Met, March:13–16, 1979.
14. AK SenGupta, Y Zhu. AIChE J 38(1):153, 1992.
15. J Melling, DW West. Proceedings of the 4th International Conference on Ion Exchange Technology at the University of Cambridge, July 1984, pp 724–735.
16. J Boegel, D Clifford. EPA Project Summary. Report No. EPA/600/S2-86/031, 1985.
17. JE Maneval, G Klein, J Sinkovic. EPA Project Summary. Report No. EPA/600/S2-85/074, 1985.
18. HJ Brauch, PhD, dissertation, University of Karlsruhe, Germany, 1984.
19. D Clifford. In: FW Pontius, ed., Water Quality and Treatment. New York: McGraw-Hill, 1990, p 561.
20. A Ramana, MS. thesis, Lehigh University, Bethlehem, PA, 1990.
21. G Boari, L Liberti, R Passino. Wat Res 10:421–428, 1976.
22. L Liberti, G Boari, R Passino. Wat Res 11:517–523, 1976.
23. D Zhao, AK SenGupta. Water Res 32:1613–1623, 1998.
24. Y Zhu, AK SenGupta. Environ Sci Technol 26(10):1990–1997, 1992.
25. A Ramana, AK SenGupta. Environ Eng Div J ASCE 118:5,755–775, 1992.
26. L Lloyd, RB Dean. J WPCF 42(5):R161–R172, 1970.
27. R Eliassen, G Tchobanoglous. J WPCE 40(5):R171–R180, 1968.
28. SJ Kang, RIM-NUT Demonstration Project. U.S. EPA Grant No. R-005858-01. Ann Arbor, MI: McNamee, Porter and Seeley, 1990.
29. K Urano, H Tachikawa. Ind Eng Chem Res 30(8):1897–1899, 1991.
30. Water Environment Research Foundation. Research Needs for Nutrient Removal from Wastewater. Project 92-WAR-1, 1994.
31. JB Nesbitt. J WPCF 41(5):701–713, 1969.
32. FX Pollio R Kunin. Environ Sci Technol 2(1):54–61, 1968.
33. J Chudoba. Wat Res 19(1):37–43, 1985.
34. M Rebhun, J Manka. Environ Sci Technol 5:606–609, 1971.
35. BE Rittman et al. Wat Sci Technol 19:517–528, 1987.
36. D Zhao. PhD dissertation, Lehigh University, Bethlehem, PA, 1997.
37. D Zhao, AK SenGupta, L Stewart. Ind Eng Chem Res 37:11,4383, 1998.
38. AK SenGupta, Y Zhu. Ind Eng Chem Res 33:2,382–386, 1994.
39. Y Zhu. PhD dissertation, Lehigh University, Bethlehem, PA, 1992.
40. W Worthy. Chem Eng News Dec. 2:27, 1991.
41. D Zhao, AK SenGupta. Ind Eng Chem Res 39:2,455, 2000.

7

Imprinted Metal-Selective Ion Exchanger

Masahiro Goto

Kyushu University, Fukuoka, Japan

I. INTRODUCTION

It is known that adsorption techniques are useful for separation and concentration of a target metal ion from a dilute solution [1]. The adsorption technique has many advantages, such as high selectivity, simple operation, and wide application. The adsorbent is crucially influential in the adsorption operation, therefore, the choice of the adsorbent often decides the success of the metal separation process.

The selectivity of adsorbents toward a target metal ion depends on the original disposition derived from functional groups introduced to the adsorbents. The functional groups often contain nitrogen, oxygen, and/or sulfur atoms, which are usually electron donor elements. The combination of these elements enables us to design a large variety of coordination groups. Nevertheless, practical adsorbents for metal separation are not so common because of the lack of innovation and the cost restriction of synthesizing novel commercial adsorbents. Regarding the circulation of metal resources, although the adsorption technique is considered to be the first choice for recovery, the selection of adsorbents is still limited.

259

Therefore, it appears desirable to develop a novel methodology to create a high-performance adsorbent for a target metal ion.

Recently, molecular recognition has drawn much attention to separation chemistry. One of the promising approaches to create a highly selective adsorbent is the "molecular imprinting technique." This concept was first proposed by Wulff and Sarhan in the early 1970s [2]. This epoch-making methodology does not require a precise molecular design and multistep procedure for producing high-performance adsorbents. We applied the molecular imprinting technique to construct a novel metal-selective ion exchanger that shows high selectivity toward the imprinted target metal ion.

II. CONCEPT OF THE MOLECULAR IMPRINTING TECHNIQUE

Molecular imprinting is a technique for preparing polymeric materials that are capable of high molecular recognition. Figure 1 illustrates the concept of the conventional molecular imprinting technique. The preparation approach to preparing imprinted polymers involves interactive pre-

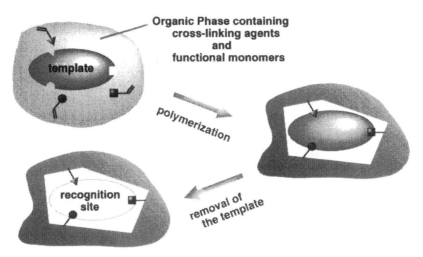

Figure 1 Concept of conventional molecular imprinting technique.

organization of functional monomers such that specific chemical interactions occur between functional monomers and "print" molecules, followed by polymerization in the presence of excess crosslinking agents. The resulting polymer contains specific binding sites that recognize the shape and the size of print molecules, and such a polymer exhibits a high selectivity for rebinding the print with which it was prepared. Because the print molecule itself directs the organization of the functional groups, specific knowledge of the imprinted structure is not necessary. The great advantage of this self-assembly approach to form specific recognition sites stems from its generality and relative simplicity. Wulff [3] demonstrated this approach by designing a polymeric receptor which utilized reversible covalent bonding for the monomer-print interactions. Molecular imprinting has now been successfully applied to create the recognition sites which employ noncovalent interactions, such as hydrogen bonding and/ or electrostatic interactions, in synthetic polymers for a variety of molecules: e.g., nucleotide bases [4], steroid hormones [5], dipeptide derivatives [6], and amino acids [7–10].

Nishide and Tsuchida first prepared metal-imprinted polymers for Cu(II), Fe(II), Co(II), and Zn(II) [11]. The imprinted polymers were constructed by using poly(4-vinyl pyridine) and 1,4-dibromobutane as the functional monomer and the crosslinking monomer, respectively (Fig. 2). The metal-imprinted polymers were confirmed to exhibit a high selec-

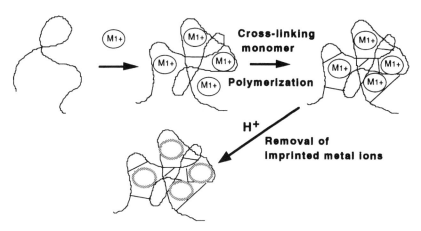

Figure 2 Schematic for the preparation of metal-imprinted ion exchanger.

tivity toward the imprinted metal ions in a batch test. Efendiev and Kabanov designed a metal-imprinted ion exchanger by introducing vinyl phosphonic acid monomer into the matrix, and Cu(II), Co(II), and Ni(II) were imprinted [12]. A high-selective ion exchanger was prepared for Cu(II) by combining two functional monomers: acrylic acid and vinyl pyridine monomers. The Cu(II)-imprinted resin provided extremely high selectivity (10- to 200-fold) compared to that of the control (unimprinted) resin [13].

Although these imprinting techniques are conceptually attractive, few practically useful adsorbents have been reported. Problems encountered include difficulty in stripping the imprinted metal ions, loss of selectivity with time, and slow rebinding kinetics. Common fundamental problems with these conventional imprinting techniques stem from recognition sites that are usually formed within the hydrophobic polymer matrix. Particularly, difficulties arise when handling water-soluble substances such as metal ions and biological components. To overcome the problems, we proposed a novel molecular imprinting technique, called "surface molecular imprinting."

III. THE SURFACE MOLECULAR IMPRINTING TECHNIQUE

The surface molecular imprinting technique is divided into two categories by the use of either an oil-in-water (O/W) or a water-in-oil (W/O) emulsion. Yu et al. and Tsukagoshi et al. first proposed the novel methodology of the surface imprinting technique to construct recognition sites on the polymer surface by utilizing the outer surfaces of O/W emulsions [14–16]. Whereas we invented the surface imprinting technique utilizing inner surfaces of W/O emulsions [17–26]. The general idea of the surface molecular imprinting technique using W/O emulsions is illustrated in Fig. 3. Surface-imprinted polymers are prepared by emulsion polymerization using a functional host molecule, an emulsion stabilizer, a polymer matrix-forming monomer, and a print molecule. The aqueous-organic interface in the W/O emulsions is functionalized as the recognizing field for a target metal ion. The target metal ion forms a complex with the functional host molecule at the inner surfaces of W/O emulsions. The orientation of the

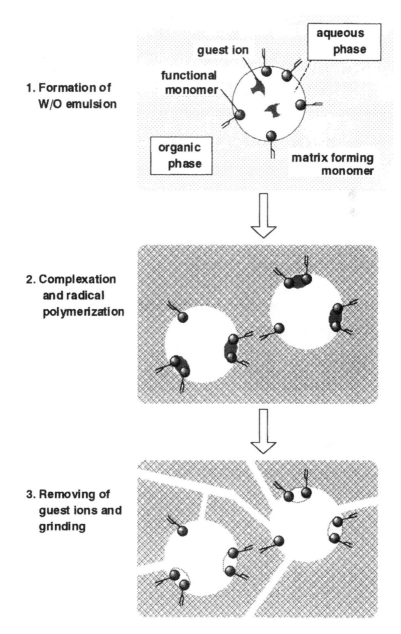

Figure 3 Schematic of surface template polymerization with W/O emulsions.

functional host molecule is controlled by the complexation with the target metal ion at the interface. The complexes between the functional host molecules and the imprinting metal ion are desired not to be totally hydrophobic nor hydrophilic, and the functional host molecules, which have an amphiphilic property, are required to achieve a high template effect for the target molecules [22]. The organic phase containing the crosslinking agent is polymerized so that the target-selective cavities are created on the polymer surfaces, not inside the polymer matrixes. This surface imprinting technique allows us to overcome the following fundamental drawbacks that stem from the formation of recognition sites within the hydrophobic polymer matrix: (1) rebinding kinetics is relatively slow and (2) it is difficult to handle water-soluble templates such as metal ions and biological components. The benefit of the surface imprinting technique eventually enables a rapid and reversible complexation of the target molecule toward the surface-imprinted polymer.

A. The Metal-Imprinted Ion Exchanger

As a typical example, the preparation scheme for a Zn(II)-imprinted polymer by the surface imprinting technique is shown in Fig. 4. The organic solution contains divinylbenzene (DVB), an extractant, as the host molecule, and a surfactant to stabilize the emulsion. Figure 5 shows the structure of typical extractants and a surfactant. Although the detailed conditions required for the host molecule are described later, extractants used in a solvent extraction process are applicable as the host molecule for a target metal ion. A high interfacial activity is required for the surfactant molecule to stabilize the emulsion at the minimum concentration. Furthermore, to form the W/O emulsion, a hydrophobic property is required for the surfactant. To satisfy these conditions, ribitol dioleyl glutamate ($2C_{18}\Delta^9GE$) was chosen as the emulsion stabilizer. A typical preparation method of a surface-imprinted ion exchanger is as follows [17,20].

2.9 g (4.8×10^{-3} mol) of dioleyl phosphoric acid (DOLPA) and 0.25 g (3.0×10^{-4} mol) of $2C_{18}\Delta^9GE$ were dissolved in 60 mL of toluene/DVB, 1/2 (v/v). The aqueous solution contained a target zinc ion, and the pH value was adjusted to a desired pH value where complete extraction is achieved. A 30 mL aqueous solution of 1.0×10^{-2} mol/L Zn(II), which

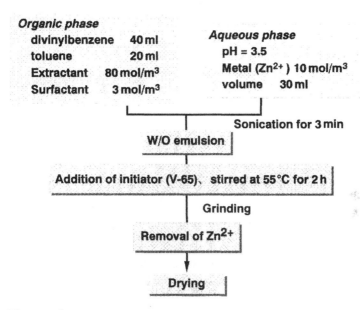

Figure 4 Flowchart of preparation method of surface metal-imprinted adsorbent.

Figure 5 Structure of extractants and surfactant used in the metal-imprinted polymerization.

was buffered with acetic acid–sodium acetate and was maintained at pH 3.5, was added, and the mixture was sonicated for 3 min to provide a W/O emulsion. Subsequently, by the addition of 0.36 g (1.4 $\times$ 10^{-3} mol) of 2,2′-azobis(2,4-dimethylvaleronitrile) [1 wt% with respect to DVB (36.56 g, 2.8 $\times$ 10^{-1} mol)], polymerization was carried out at 55°C for 2 h in a stream of nitrogen. The bulk polymer was dried under vacuum and ground into particles. Finally, the particles were washed with 1.0 mol/ L hydrochloric acid solution to remove Zn(II) and then filtered. This procedure was repeated until Zn(II) in the filtrate became negligible. The Zn(II)-imprinted polymer was dried in vacuum. Figure 6 shows a typical view of the Zn(II)-imprinted polymer prepared from W/O emulsions by scanning electron microscopy (SEM). Substantial traces of aqueous phases in the emulsions are observed in the polymer. The recognition sites for the target zinc ions are constructed on the surfaces of the inner cavities of the polymer. A number of micropores in the polymers facilitate diffusion of the target metal ion into the polymer. According to the N$_2$ adsorption test, the specific surface area of the imprinted polymer was found to be 15 m^2 per gram of polymer, and the average pore size was around 10– 20 nm.

B. Key Factors for Preparing a High- Performance Ion Exchanger

For preparing highly selective surface-imprinted polymers, it is necessary to fix the binding sites for a target molecule rigidly on the polymer surface. The assessment of a suitable functional host molecule, which forms binding sites and recognizes the target molecule, is a matter of great importance. At the beginning we considered the requirement of functional host molecules with strong binding ability for the target molecule and chose two typical extractants: di-2-ethylhexyl phosphoric acid (D2EHPA) and dioleyl phosphoric acid, which are used in solvent extraction operations. We prepared a Zn(II)-imprinted adsorbent by surface template polymerization using each extractant and investigated the adsorption performance. Figure 7 shows the adsorption behavior of the imprinted and unimprinted polymers for zinc ions. Although these extractants have a similar extraction ability for zinc ions in a solvent extraction test, the two imprinted adsorbents exhibited quite different adsorption behavior.

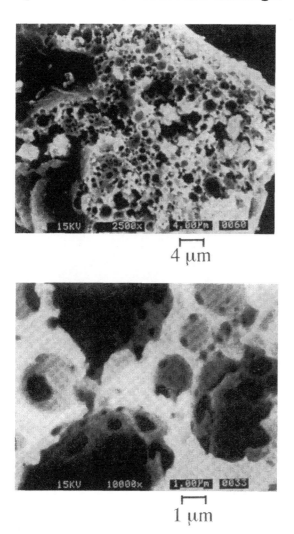

4 μm

1 μm

Figure 6 SEM photograph of metal-imprinted polymer.

Here it should be noted that the interfacial activity of the extractants differs significantly. DOLPA has a phosphoric acid group in the hydrophilic part and two long oleyl chains in the hydrophobic part; both the binding ability for Zn(II) and the interfacial activity of DOLPA are very high. In the preparation of metal-imprinted polymers, DOLPA molecules

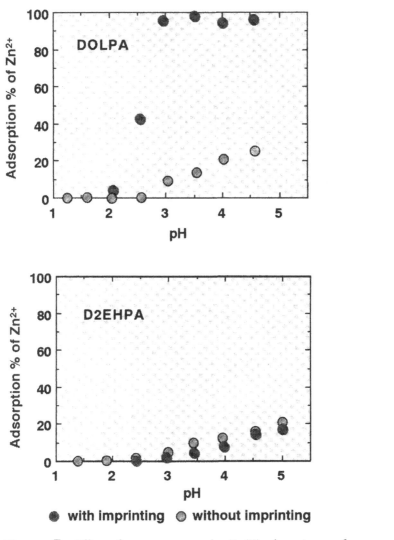

Figure 7 Effect of extractants on the Zn(II)-adsorption performance of metal-imprinted polymers with D2EHPA and DOLPA.

are arranged at the surface of polymers due to the high interfacial activity. While D2EHPA molecules, which do not have a high interfacial activity, are deduced to be buried into the polymer matrix during polymerization. An important factor in making a good metal-imprinted adsorbent was found to be a high interfacial activity of the extractant used as the host molecule for a target metal ion.

The adsorbent prepared by DOLPA exhibited a high-imprinting effect as far as the amount of metal ions adsorbed is concerned. However, the Zn(II)-imprinted polymer showed relatively low selectivity for Zn(II) in the competitive Cu(II)–Zn(II) adsorption from an aqueous solution. This is probably due to insufficient rigidity of the polymer matrix, which causes increased swelling of the imprinted polymers. In comparison, divinylbenzene polymers without DOLPA provided a more rigid matrix, and the swelling of the polymer was suppressed. Thus, the structure of the functional host molecule has a dominant effect on the rigidity of the polymeric matrix. It has been established that host molecules that possess long alkyl chains in their hydrophobic part affect the matrix rigidity. Fixing the imprinted host molecules rigidly on the polymer surface proved to be essential in ensuring a high selectivity toward a target metal ion.

To fix the recognition sites more rigidly and create stronger interactions between the functional host molecules and imprint molecules, we tried to make Zn(II)-imprinted polymers by the following two approaches: (1) postirradiation with gamma rays to make the polymer matrix more rigid and (2) the design of novel functional host molecules for molecularly imprinted adsorbents.

1. Postirradiation with Gamma Rays

We found that rigidity of the imprinted polymers is one of the important factors in producing a high-selective ion exchanger. Therefore, we considered the use of gamma rays to make the matrix more rigid. A Zn(II)-imprinted polymer was prepared with DOLPA and DVB as a functional host molecule and a polymer matrix-forming monomer. After drying in vacuo, the imprinted polymer was subjected to irradiation by gamma rays to bind DOLPA strongly and to crosslink the polymer matrix [19]. The matrix was made rigid by the irradiation.

Figure 8 shows the pH dependence for competitive adsorption of Zn(II) and Cu(II) by the Zn(II)-imprinted polymer prepared with

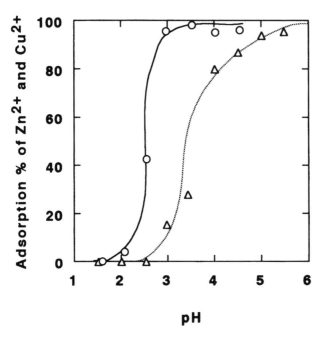

Figure 8 pH dependence for adsorption of Zn(II) (circle) and Cu(II) (triangle) with a Zn(II)-imprinted polymer prepared with DOLPA without gamma ray irradiation.

DOLPA without gamma ray irradiation after the initial polymerization. The amounts of metal ions in the rebinding test were set below the maximum capacity of the Zn(II)-imprinted sites (1.1×10^{-6} mol for 0.10 g of polymer) to evaluate the binding affinity of the polymer. The Zn(II)-imprinted polymer adsorbed both Zn(II) and Cu(II) over the entire pH range. However, this imprinted polymer had less selectivity for adsorption of Zn(II) over Cu(II).

In a solvent extraction system (Fig. 9), only low selectivity for Zn(II) is observed because the functional molecule (extractant) can assume both a tetrahedral configuration for Zn(II) and a square planar configuration for Cu(II) due to its flexibility in the organic solvent. It is noted that the Zn(II) selectivities of the Zn(II)-imprinted polymer prepared with DOLPA without gamma ray irradiation (Fig. 8) and in solvent extraction

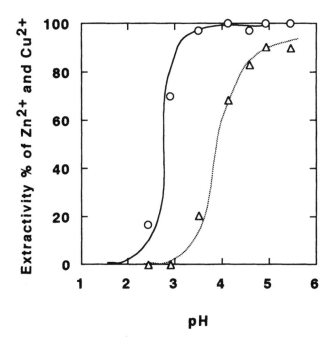

Figure 9 pH dependence for solvent extraction of Zn(II) (circle) and Cu(II) (triangle) from aqueous solution into toluene containing DOLPA.

by DOLPA (Fig. 9) are very similar. Thus, the Zn(II)-imprinted polymer has a metal ion selectivity equal only to that of the functional molecule, as is the case for other conventional imprinted polymers [11–13].

Figure 10 shows the pH dependence of adsorption of Zn(II) and Cu(II) by the Zn(II)-imprinted polymers with DOLPA. No difference is noted for Zn(II) adsorption by the imprinted polymers with and without postirradiation by gamma rays. An irradiated blank polymer, which did not include DOLPA, adsorbed no metal ions (data not shown). These results clearly show that irradiation of gamma rays does not destroy the Zn(II)–DOLPA complex and produces few functional groups, such as carbonyl groups. In contrast, the Cu(II) binding ability of the imprinted/ gamma ray irradiated polymer was markedly decreased. Thus the imprinted/gamma ray irradiated polymers can distinguish the coordination of Zn(II) from that of Cu(II).

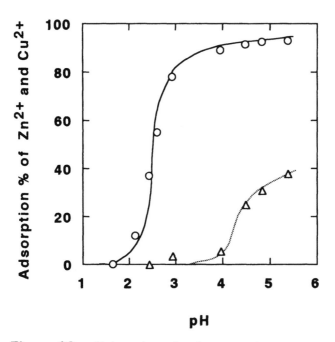

Figure 10   pH dependence for adsorption of Zn(II) (circle) and Cu(II) (triangle) by a Zn(II)-imprinted polymer prepared from DOLPA with gamma ray irradiation.

From the swelling percentage in THF for all of the polymers (Table 1), the irradiated polymers were found to swell less than nonirradiated polymers. The ratio of DOLPA released from each polymer matrix during acid washing operations was remarkably reduced by irradiation with gamma rays (Table 1). Improvement in the selectivity of imprinted/ gamma ray irradiated polymers is attributed to the induced crosslinking in the polymer matrix by irradiation that renders the polymer matrix rigid, and thereby enhances the stability of the binding sites toward recognition of Zn(II). Furthermore, the double bond in the oleyl chains of DOLPA can form single bonds with free radicals on each end to combine DOLPA rigidly with the polymer matrix.

These results show that rigid and dimensionally stable metal ion– imprinted polymers can be prepared, which recognize metal coordination by anchoring the functional host molecule, such as DOLPA, to the poly-

Table 1 Swelling Ratio of Polymers in THF

Polymers	Swelling ratio (%)	Released extractant (%)
Unimprinted polymer	27	1.5
Gamma ray irradiated unimprinted polymer	22	1.3
Imprinted polymer	37	3.1
Gamma ray irradiated imprinted polymer	27	1.5

mer surface. It is known that properties such as matrix rigidity bring about poor mass transfer in conventional imprinted polymers because recognition sites exist inside the polymer matrix. Whereas, in the surface imprinting technique, adsorption sites are formed on the polymer surfaces. The ability to enhance the matrix rigidity without decreasing mass transfer is therefore an important advantage in surface template polymerization. Furthermore, the combination of the surface template polymerization with irradiation of gamma rays offers a potential technique to construct highly selective, molecule-recognizing polymers which are applicable to the adsorption of various metal ions.

2. Design of Novel Functional Host Molecules

In order to fix the recognition sites more rigidly, we reconsidered design requirements for the functional host molecules. The following qualities were specified: (1) strong binding ability for the target metal ions, (2) high interfacial activity, and (3) no detrimental effect on the polymer matrix. To fulfill these conditions, we designed the functional host DDDPA (1,12-dodecanediol-O, O'-diphenylphosphonic acid), which has two phosphonic acid groups and two benzene rings in the molecular structure (Fig. 5) [18,22,27]. The two phosphonic acid groups were prepared to produce strong binding for Zn(II), the target metal ion. The two benzene rings were included in the hydrophobic portion of DDDPA to preserve the polymer matrix, which is formed by divinylbenzene. Alkyl chains, which link the two phenylphosphonic acid units, controlled the interfacial activity of the functional host molecules and their solubility in organic solvents. Zn(II)-imprinted polymers derived from this multifunctional host are expected to be highly selective adsorbents toward Zn(II) over Cu(II) because the polymers possess both the rigid polymer matrixes and the strong binding ability due to the specificity of the multifunctional host molecule.

The interfacial activity of DDDPA is eight times greater than that of DOLPA [27]. Therefore, DDDPA satisfied the second requirement in the design of functional hosts. Figure 11 shows the pH dependence for adsorption of Zn(II) and Cu(II) on a Zn(II)-imprinted polymer prepared with DDDPA. The percent adsorption was enhanced with increased pH for both ions. However, the imprinted polymer adsorbed Zn(II) much more effectively than it adsorbed Cu(II) over the entire pH range. The ability of the imprinted polymers to adsorb Zn(II) is significantly higher than that for Cu(II) because of the strong interaction between Zn(II) and DDDPA. This property of DDDPA satisfied the first requirement for functional molecules applicable in molecular imprinting techniques. It

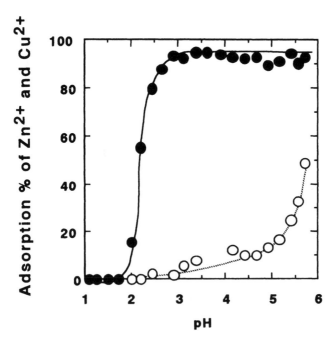

Figure 11   pH dependence of the adsorption of Zn(II) (closed circle) and Cu(II) (open circle) by a Zn(II)-imprinted polymer using DDDPA.

should be noted that Zn(II) is completely separated from Cu(II) in aqueous solutions with pH ca. 3. This high selectivity is caused by the Zn(II)-imprinted cavity formed on the surface of polymers.

We have prepared many of metal ion–imprinted polymers using various organophosphorus extractants as a functional molecule. However, we could not always obtain an excellent template effect due to the difficulty in fixing the recognition site on the polymer surface sufficiently. Thus, we have further investigated in more detail the effect of molecular structure of bifunctional organophosphorus extractants (BDP, ODP, and DDDPA), as shown in Fig. 5, each of which having a different length of spacer ($-(CH_2)n-$; $n = 4, 8, 12$) linking two phosphonic acid groups and two aromatic rings. The two phosphonic acid groups are expected to create a high interaction with imprinting metal ions because the phosphonic group is known to have a high affinity for a wide variety of metal ions. The two aromatic rings play an important role in fixing the recognition sites rigidly by specific interaction such as $\pi-\pi$ stacking with the polymer matrix that is formed by DVB. The interfacial activity at the oil–water interface depends on the length of alkyl chains linking the two phosphonic acid groups. Thus, the Zn(II)-imprinted polymers with DDDPA, ODP, and BDP, which have different lengths of alkyl chains, were prepared to investigate the contribution of the straight alkyl spacer on the template effect. The Zn(II)-imprinted polymer with DDDPA exhibited a high adsorption ability compared to the other polymers. However, the Zn(II)-imprinted polymers with ODP and BDP, whose structures are similar to that of DDDPA, hardly exhibited a high adsorption ability.

In the liquid–liquid extraction system, DDDPA and ODP show a similar tendency for the extraction due to the high affinity of phosphonic group to zinc ions [27], while BDP hardly extracted zinc ions because of the low solubility to the organic solvent. Therefore, it is found that the bifunctional molecule BDP is not suitable for a recognition host in the surface template polymerization technique. A surprising result is that although the liquid–liquid extraction system using ODP exhibited an excellent extraction profile for zinc ions, the Zn(II)-imprinted polymer with the same extractant ODP did not show a good adsorption performance for zinc ions. In addition, the adsorption of the ODP polymer was saturated at higher pH range (Fig. 12). It is considered that the arrangement

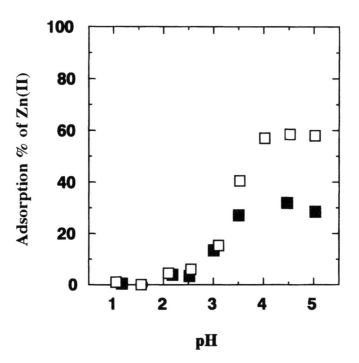

Figure 12 pH dependence on zinc ion adsorption by a Zn(II)-imprinted (open square) and unimprinted (closed square) polymer using ODP.

of the extractant ODP on the polymer surface is not sufficient to construct the recognition sites for zinc ions. Hence, we found that the length of the alkyl spacer has a dominant effect to imprint a target metal ion on the surface of polymers. In this respect, introduction of a bifunctional monomer having a straight twelve methylene spacer, DDDPA, proved to be exactly appropriate in the surface template polymerization technique.

The interfacial activity of functional host molecules is one of the most important factors in arranging the recognition sites on the polymer surface when preparing metal-imprinted polymers. This is also related to the creation of rigid recognition sites and facilitating high interactions between a functional host molecule and an imprint molecule on the surface of polymers. The adsorption equilibrium of the functional molecules is expressed as follows:

$$F_{org} = F_{ad} : K_F \qquad (1)$$

where K_F is the interfacial adsorption equilibrium constant, while F and the subscript ad denote the functional host molecule and the adsorption state at the interface, respectively. The relationship between interfacial tension and the amount of functional molecule adsorbed at the interface is expressed by the Gibb's adsorption equation. Assuming a Langmuir adsorption isotherm between the amount of functional host molecule adsorbed and the bulk concentration of functional molecule, C_F, the relationship between interfacial tension, γ, and C_F at temperature T is expressed as follows:

$$\gamma = \gamma_0 - \frac{RT}{S_F} \ln (1 + K_F C_F) \qquad (2)$$

where γ_0 is the interfacial tension between the organic solvent and the aqueous solution, R is the gas constant, and S_F is the interfacial area occupied by a unit mole of functional host molecule. The values of K_F and S_F can be obtained from the experimental results for the interfacial tension and Eq. (2) by using a nonlinear regression method. The interfacial adsorption equilibrium constants for the functional host molecules are listed in Table 2. In particular, the interfacial activity of DDDPA is much higher than that of the other host molecules. Based on the results it is found that the DDDPA satisfied the first requirement as a criterion in the design of functional molecules and efficiently constructed the recognition sites for zinc ions on the polymer surface. A phosphonic extractant having a

Table 2 Interfacial Adsorption Equilibrium Constants of Various Functional Molecules in Toluene

Functional monomers	Interfacial adsorption equilibrium constant K_F (m³/mol)
DDDPA	307
ODP	158
BDP	60
D2EHPA	2.1
DOLPA	48

suitable alkyl chain is one of the best candidates to make recognition sites for a target metal ion using the surface template polymerization technique.

On the basis of our design guideline, we also synthesized several other functional host molecules. We then determined the desirable ranges for the swelling ratio (rigidity of recognition sites) of Zn(II)-imprinted polymers and the interfacial activity of functional host molecules (see Fig. 13) [22]. The functional host molecules that have aromatic rings and a high interfacial activity, such as DDDPA and n-DDP (n-dodecylphosphonic acid), were able to fix the recognition sites rigidly on the polymer surface and, consequently, produced a high selectivity toward zinc ions. If only the functional hosts exhibited appropriate interfacial activity and strong binding characteristics, then the polymer would have demonstrated an excellent template effect by being irradiated with gamma rays, as in the case of DOLPA.

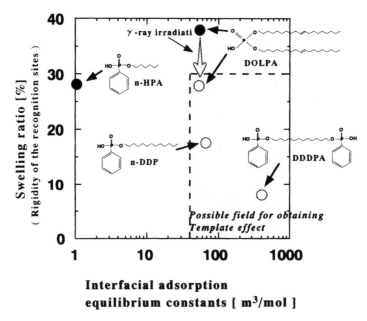

Figure 13 Essential properties for functional host molecules. The hatched region is suitable for molecular imprinting. (From Ref. 22.)

IV. APPLICATION TO THE SEPARATION OF LANTHANIDE IONS

Lanthanide elements are in great demand for the production of novel advanced materials used in various electronic, optical, and magnetic devices. An efficient process for separating lanthanide elements is still under study because these elements behave almost identically due to the similarity of chemical and physical properties among the lanthanides. Currently, the simplest, most efficient process for separating lanthanides is column separation using an appropriate stationary phase. The advantage of this process is that it produces a high concentration of lanthanides. However, an expensive chelating reagent is required as a selective eluent because a highly selective cation exchanger has not been developed as a stationary phase. Therefore, we have tried to prepare highly selective ion exchangers toward lanthanide elements by using surface template polymerization.

Most of the organophosphate–lanthanide(III) complexes have a nona-coordination structure, similar to the tricapped trigonal prism coordination in the lanthanide(III) series [28]. Thus, size recognition is required to prepare the surface-templated resins for the effective separation of lanthanide elements.

We prepared two different Nd(III)-imprinted polymers by surface template polymerization using phenylphosphonic acid monododecyl ester (abbreviated with n-DDP) or 2-ethylhexyl phosphonic acid mono-2-ethylhexyl ester (commercial name, PC-88A) as functional host molecules. The binding ability and selectivity of these polymers for Nd(III) and La(III) ions were evaluated by a batchwise method. We investigated the criteria for the optimum structure of functional molecules in lanthanide separation [29].

We know that the binding affinity between the functional molecule and the target metal ion is one of the essential factors in preparing highly efficient imprinted resins. Therefore, first, the degree of binding affinity of the functional molecule to Nd(III) and La(III) was evaluated by the liquid–liquid extraction system. Figures 14(a) and (b) show extraction behavior of Nd(III) and La(III) with each functional molecule. Although n-DDP exhibited a high affinity for both Nd(III) and La(III), effective separation was not observed. On the other hand, PC-88A has relatively

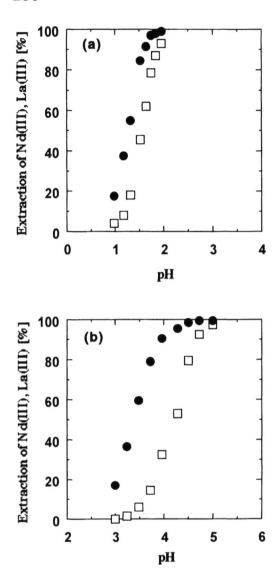

Figure 14 (a) Extraction behavior of Nd(III) (closed circle) and La(III) (open square) ions using n-DDP as an extractant. (b) Extraction behavior of Nd(III) (closed circle) and La(III) (open square) ions using PC-88A as an extractant.

good affinity and selectivity toward Nd(III) over La(III). In an industrial extraction process, PC-88A is known to be an excellent extractant for the separation of lanthanide ions. Based on the above results, we found that both functional molecules possess adequate affinity to the target metal ion, Nd(III).

Figures 15(a) and (b) show adsorption behavior for Nd(III) and La(III) on the Nd(III)-imprinted resins using n-DDP or PC-88A, respectively. The Nd(III)-imprinted polymer with n-DDP exhibited high selectivity to Nd(III) over La(III) in a wide pH range, while such a high selectivity was not observed in the liquid–liquid extraction system with n-DDP as the extractant. This result suggests that the functional host molecule, n-DDP, in the Nd(III)-imprinted polymer is fixed in the conformation that is complementary to the target Nd(III). On the other hand, the Nd(III)-imprinted polymer using PC-88A showed inferior adsorption ability and selectivity for both Nd(III) and La(III) compared to the Nd(III)-imprinted polymer using n-DDP, although PC-88A had a high binding affinity to Nd(III) in the liquid–liquid extraction system.

Effective ion exchange capacity and adsorption equilibrium constant were evaluated to characterize the imprint effect for Nd(III) on the Nd(III)-imprinted polymers. The effective ion exchange capacity for the Nd(III)-imprinted polymers with n-DDP is much higher than that with PC-88A, which suggests that more accessible binding sites were constructed on the Nd(III)-imprinted polymer surface when n-DDP was used as a functional molecule. Furthermore, it was clarified that the adsorption equilibrium constant of the Nd(III)-imprinted polymer with n-DDP was about 100 times higher than that with PC-88A. These results indicated that the recognition sites with high binding affinity toward Nd(III) were created on the Nd(III)-imprinted polymer by using n-DDP as a functional molecule.

An interesting question is why the structural difference in n-DDP and PC-88A causes such a different adsorption behavior. PC-88A and n-DDP are phosphonic acid derivatives, however, PC-88A has two branched alkyl chains, 2-ethylhexyl groups, whereas n-DDP has an aromatic ring and a dodecyl group. On the basis of the criteria shown in Fig. 13, n-DDP becomes a candidate satisfying the above requirements (interfacial adsorption equilibrium constant, $K_F = 67$ m^3/mol) [22], therefore binding sites are fixed rigidly on the polymer surface. In contrast, PC-88A has

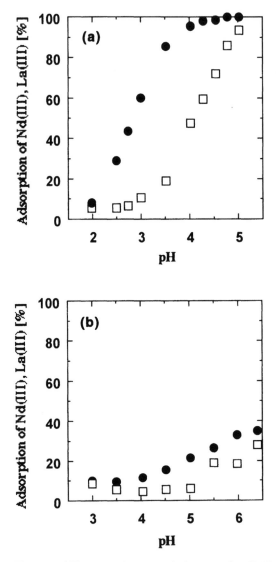

Figure 15 (a) Adsorption behavior of Nd(III) (closed circle) and La(III) (open square) ions on the Nd(III)-imprinted polymer with n-DDP. (b) Adsorption behavior of Nd(III) (closed circle) and La(III) (open square) ions on the Nd(III)-imprinted polymer with PC-88A.

no aromatic rings and little interfacial activity ($K_F = 3.3$ m^3/mol). Based on this knowledge, binding sites are buried in the polymer matrix, which results in the reduction of adsorption ability.

This is the first report to prepare lanthanide-imprinted resins. Also, we found that the essential factors in the lanthanide-imprinted resins are similar to those in the Cu(II)- and Zn(II)-imprinted resins to create highly efficient adsorbents.

V. PREPARATION OF METAL-IMPRINTED MICROSPHERES

As described before, we prepared a novel functional host molecule, 1,12-dodecanediol-O, O'-diphenylphosphonic acid (DDDPA), which proved to be suitable for preparing a metal-imprinted ion exchanger possessing a high selectivity to zinc ions. The rigidity of the polymer matrix was another vital factor to produce a template effect for the target ion. Therefore, the matrix-forming monomer trimethylolpropane trimethacrylate (TRIM), which has three polymerizable groups, was introduced to make the polymer matrix more rigid. The resulting recognition sites on the polymer surface were more well defined and of better quality, which was reflected in the higher selectivity. The recognition capabilities of the metal-imprinted resins prepared with TRIM were evaluated by a selective adsorption test of zinc over copper ions. The results revealed that the combination of DDDPA and the cross linking agent TRIM serves to align the recognition sites with an exclusive template effect constructed on the polymer surface.

The molecular imprinting technique is also useful for developing novel microspheres used in a column chromatographic operation. Therefore, we prepared Zn(II)-imprinted macrospheres with DDDPA, which was spherically well defined and uniform, by using the surface template polymerization technique with water-in-oil-in-water (W/O/W) multiple emulsions. The obtained polymer beads can be applied as it is as a metal adsorbent without either grinding or sieving. In the basic fundamental research, we succeeded in forming a stable microspheric polymer using DVB. Then, we tried to add highly selective recognition sites to the surface of the microspheric polymers by utilizing the surface template polymerization technique. We focused on TRIM, which has three polymerizable groups to make the polymer-matrix more rigid. The TRIM offers better prospects of preparing macroporous spheres in a single step and forming superior metal-recognition sites. The recognition abilities of the Zn(II)-imprinted spherical polymer were systematically investigated along with those of the polymers prepared by W/O emulsions.

Prior to use as the column adsorbent, batchwise adsorption tests were conducted by using the zinc-imprinted and unimprinted microspheres to make an assessment on the effectiveness of the imprinting technique. Then, the separation of zinc ions was carried out by an adsorption column packed with the zinc-imprinted microsphere or a commercial chelating resin which has a similar phosphoric group.

A. Preparation of Zn(II)-Imprinted Microspheres by Multiple Emulsions

A metal ion–imprinted microsphere can be prepared by the surface molecular imprinting technique using a multiple W/O/W emulsion. A 40 cm^3 TRIM solution containing 60 mM DDDPA and 20 mM 2C$_{18}\Delta^9$GE, was mixed with 20 cm^3 toluene containing 5 vol% 2-ethylhexyl alcohol as a solubility modifier. An aqueous solution (30 cm^3) of 10 mM Zn(NO$_3$)$_2$, which was buffered with 100 mM acetic acid–sodium acetate and maintained at pH 3.5, was added to the organic solution. The mixture was sonicated for 4 min to form a stable W/O emulsion. The W/O emulsion was put into a 500 cm^3 aqueous phase containing 15 mM sodium dodecyl sulphate (SDS) and 10 mM Mg(NO$_3$)$_2$ as an ionic strength controller (pH 3.5, buffered with 100 mM acetic acid–sodium acetate) stirring at the rate of 500 rpm, then multiple W/O/W emulsions were formed. The polymerization was performed according to the procedure previously prescribed [24]. An unimprinted microsphere was also prepared without zinc ions as a control microsphere.

B. Characterization of the Zn(II)-Imprinted Microsphere

Highly crosslinked Zn(II)-imprinted microspheres were prepared by the surface template polymerization with W/O/W emulsions. The microspheres were directly obtained without grinding or sieving by using the multiple emulsions. The emulsion droplets served to form spherical particles at the single step. The Zn(II)-imprinted and unimprinted microspheres were obtained as particles of volume-averaged diameter ca. 25 μm. The size distribution was relatively narrow (15–30 μm). They were obtained in approximately 80–90% yields. Figure 16 shows a typical view of Zn(II)-imprinted microspheres by scanning electron microscopy. The adsorbents prepared by the multiple emulsions are spherical such as commercially available chelating resins. The recognition sites are formed on the surfaces of micropores inside the particles. Sodium dodecyl phosphate, used as a dispersion stabilizer, is mainly fixed on the outer surfaces of the polymer and plays an important role in suppressing the condensation of microspheres during preparation [24]. A number of micropores in the polymers facilitate diffusion of target metal ions (zinc ions) into the polymer matrix.

In the surface molecular imprinting technique, it is important to make the polymer matrix rigid enough to allow stiff attachment of the functional host molecules (recognition sites); this results in creating a high selectivity for target metal ions. We measured the swelling ratio

4 μm

Figure 16 Typical SEM photographs of imprinted microspheres prepared by W/O/W emulsions.

of the imprinted microsphere, which is an indicator of the rigidity of the polymer matrix. The lower swelling ratio means that the recognition sites are fixed more rigidly on the polymer surfaces. As shown in Fig. 13, we revealed that polymers showing less than 30% in the swelling ratio were able to produce a high template effect for a target metal [22]. The swelling ratio of the polymer matrix of the Zn-imprinted microsphere was measured to be 22%. As the swelling ratio was less than 30%, an efficient template effect was produced for the templated zinc ions [24].

C. Competitive Adsorption Behavior of Zn(II) and Cu(II) of the Zn(II)-Imprinted and Unimprinted Microspheres

Figure 17(a) and (b) show the pH dependence of the competitive adsorption for zinc and copper ions on the Zn(II)-imprinted and unimprinted microspheres, respectively. The percent adsorption increased with increased pH for all polymers. This result means that the proton dissociation from the phosphonic acid group in the functional host molecule fixed on the polymer surface plays an important role in adsorbing both metals (zinc and copper). The Zn(II)-imprinted microsphere prepared in the presence of zinc ions exhibited a high selectivity to zinc ions over copper ions in the whole pH range (from 1 to 4). On the other hand, the unimprinted microsphere, which was prepared without zinc ions, did not afford a good performance and showed similar adsorption profiles for both metal ions. These results clearly demonstrated that the high selectivity was produced by the imprinting technique, by which the functional molecule DDDPA was suitably implanted on the polymer surfaces. A desirable coordination space (tetrahedral configuration) is formed on the surfaces to facilitate the interaction with the imprinted zinc ions. The recognition sites memorized on the surfaces should have high affinity to zinc ions excluding the accommodation of copper ions (square planar configuration). On the other hand, in the unimprinted microsphere, the functional host molecules were distributed randomly on the polymer surfaces, where metal ions showed similar tendencies of adsorption behavior. These results indicate that the functional host molecules (memorized cavities) on the surfaces of the

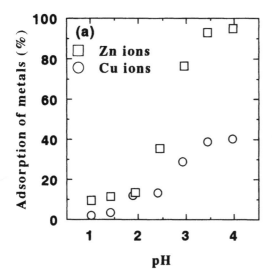

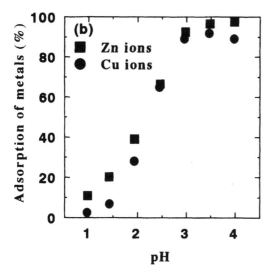

Figure 17 pH dependence of the adsorption for zinc and copper ions on the (a) Zn(II)-imprinted and (b) unimprinted microspheres.

Zn(II)-imprinted microsphere are immobilized with the strictly tetrahedral configuration suitable for zinc ions.

D. Chromatographic Separation of Zn(II) and Cu(II) by a Packed Column with Zn(II)-Imprinted Microsphere and Commercial Chelating Resin

To demonstrate the specific selectivity to the target zinc ions, breakthrough and elution tests were conducted by the adsorption column packed with the Zn(II)-imprinted microspheres or a commercially available chelating resin (Duolite C-467) possessing phosphoric groups. Figure 18 illustrates the schematic diagram of the column packed with microspheres for the separation of metal ions.

The breakthrough profiles of zinc and copper ions by the column packed with the Zn(II)-imprinted microspheres are shown in Fig. 19. The breakthrough of copper ions immediately begins in the initial stage of the operation (about 100 cm^3), while that of zinc ions begins at around 300 cm^3. An efficient and clearcut separation between the two metal ions was achieved by using the imprinted microsphere, as expected from the results of batchwise experiment as shown in Fig. 17(a). Figure 20 shows the elution profiles of zinc and copper ions from the loaded column packed with the Zn(II)-imprinted microspheres by a 1 N hydrochloric acid solution. An extremely sharp elution profile is observed at around 10 cm^3 only for zinc ions, which were concentrated as high as more than 90 times compared to that of the initial feed solution, while the elution of copper ions is not observed. Furthermore, to prove the superiority of the Zn(II)-imprinted microsphere with respect to metal selectivity, a commercially available chelating resin possessing a similar functional group, phosphoric acid, was tested as a control experiment. Figure 21 exhibits the breakthrough profiles for zinc and copper ions from the column packed with Duolite C-467. However, an efficient separation performance was not observed between the two metals. This result demonstrates that the memorized cavities, which have a high affinity to zinc ions, were realized on the surfaces of the Zn(II)-imprinted microsphere.

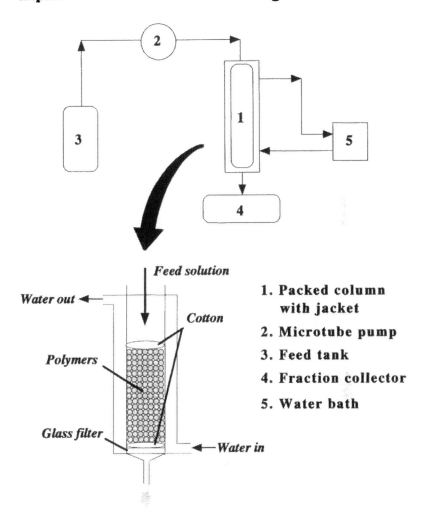

Figure 18 Schematic diagrams of the column packed with microspheres for the separation of metal ions.

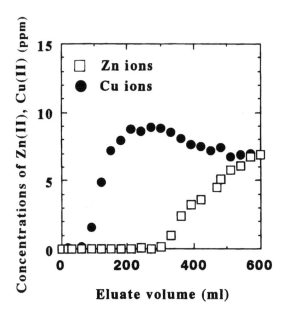

Figure 19 Breakthrough profiles for the separation of zinc and copper ions in the column packed with the Zn(II)-imprinted microspheres.

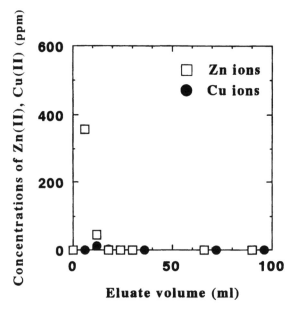

Figure 20 Elution profiles of zinc and copper ions from the Zn(II)-imprinted microspheres.

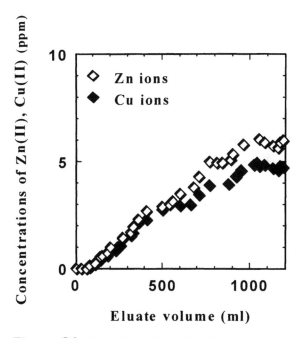

Figure 21 Breakthrough profiles for the separation of zinc and copper ions in the column packed with Duolite C-467.

VI. NEW DIRECTIONS

In addition to highly selective metal-imprinted polymers, we have also prepared an enantioselective polymer by incorporating the functional host n-DDP (Fig. 13) by surface template polymerization with W/O emulsions [10,25]. In this study, tryptophan methyl ester (TrpOMe) was chosen as a target material. The L-TrpOMe-imprinted polymer prepared in the presence of L-TrpOMe accomplished a highly characteristic binding with L-form over D-form in the entire pH range. However, the unimprinted polymer prepared similarly but in the absence of L-TrpOMe afforded no evidence of enantiomeric separation in this instance because the phosphonic groups are randomly distributed on the polymer surface. In addition, D-TrpOMe-imprinted polymer preferentially adsorbed D-form over L-form.

We are now extending this novel surface-imprinting technique to create artificial biocatalysts that mimic a variety of enzymes. Using the newly synthesized functional host molecule, oleyl imidazole, an enzyme-mimic polymer has been prepared by imprinting a substrate analog through the complex formation between a cobalt ion and the imidazole moiety [23]. The catalytic properties of artificial biocatalysts were investigated on the hydrolysis reaction of an amino acid ester by comparing to several control experiments. The imprinted polymer exhibits much higher catalytic activity compared to that of their control polymer. These results suggest that complementary specific recognition sites were constructed by the imprinting guest molecule and by the functional host molecules which are specially positioned on the polymer surface. We hope that our molecular surface-imprinting technique will find useful applications in the future.

REFERENCES

1. RW Grimshaw, CE Harland. Ion-exchange: Introduction to Theory and Practice; London: The Chemical Society Publications, 1975.
2. G Wulff, A Sarhan. Angew Chem, Int Ed 11:341–341, 1972.
3. G Wulff. CHEMTECH Nov: 19–26, 1998.
4. B Sellergren, KJ Shea. J Chromatogr, A 690:29–38, 1995.
5. SH Cheong, S McNiven, A Rachkov, R Levi, K Yano, I Karube. Macromolecules 30:1317–1322, 1997.
6. C Yu, K Mosbach. J Org Chem 62:4057–4060, 1997.
7. S Vidyasankar, M Ru, FH Arnold. J Chromatogr, A775:51–56, 1997.
8. M Yoshida, K Uezu, M Goto, S Furusaki. Chem Lett 1998:925–926.
9. M Yoshida, K Uezu, M Goto, S Furusaki. J Appl Poly Sci 78:695–703, 2000.
10. M Yoshida, Y Hatate, K Uezu, M Goto, S Furusaki. Coll Surf, 169:259–269, 2000.
11. H Nishide, E Tsuchida. Makromol Chem 177:2295–2302, 1976.
12. AA Efendiev, VA Kabanov. Pure Appl Chem 54:2077–2083, 1982.
13. W Kuchen, J Schram. Angew Chem, Int Ed 27:1695–1702, 1988.
14. KY Yu, K Tsukagoshi, M Maeda, M Takagi, Anal Sci 8:701–705, 1992.
15. K Tukagoshi, KY Yu, M Maeda, M Takagi. Bull Chem Soc Jpn 66:114–120, 1993.
16. K Tsukagoshi, KY Yu, M Maeda, M Takagi, T Miyajima, Bull Chem Soc Jpn 68:3095–3101, 1995.

17. K Uezu, H Nakamura, M Goto, M Murata, M Maeda, M Takagi, F Nakashio. J Chem Eng Jpn 27:436–439, 1994.
18. M Yoshida, K Uezu, M Goto, F Nakashio. J Chem Eng Jpn 29:174–177, 1996.
19. K Uezu, H Nakamura, J Kanno, T Sugo, M Goto, F Nakashio. Macromolecules 30:3888–3891, 1997.
20. K Uezu, M Goto, F Nakashio. ACS Symp Ser 703:278–289, 1998.
21. K Uezu, M Yoshida, M Goto, S Furusaki, CHEMTECH April:12–19, 1999.
22. M Yoshida, K Uezu, M Goto, S Furusaki. Macromolecules 32:1237–1243, 1999.
23. E Toorisaka, M Yoshida, K Uezu, M Goto, S Furusaki. Chem Lett 1999: 387–388.
24. M Yoshida, K Uezu, M Goto, S Furusaki, J Appl Poly Sci 73:1223–1230, 1999.
25. M Yoshida, Y Hatate, K Uezu, M Goto, S Furusaki. J Polym Sci, Part A: Polym Chem 38:386–394, 2000.
26. K Araki, M Yoshida, K Uezu, M Goto, S Furusaki. J Chem Eng Jpn submitted, 2000.
27. M Yoshida, K Uezu, M Goto, F Nakashio. J Poly Sci, Part A: Polym Chem 36:2727–2734, 1998.
28. P Comba, K Gloe, K Inoue, T Krüger, H Stephan, K Yoshizuka. Inorg Chem 37:3310–3316, 1998.

8

Synthesis and Characterization of a New Class of Hybrid Inorganic Sorbents for Heavy Metals Removal

Arthur D. Kney

Lafayette College, Easton, Pennsylvania

Arup K. SenGupta

Lehigh University, Bethlehem, Pennsylvania

I. INTRODUCTION

Selective removal of trace concentrations (mg/L to μg/L) of dissolved heavy metals from contaminated water and wastewater remains a challenging and a commonly encountered separation problem facing many industries and publicly owned treatment works (POTW). The use of beads or granules of chemically stable polymeric chelating exchangers in fixed-bed processes have, to a great extent, resolved the problem technically, but they are often found too expensive to justify their applications for heavy metals removal from water and wastewater. Both amorphous and crystalline forms of iron oxides, iron oxyhydroxides, and iron hydroxides have long been known to exhibit high sorption affinities toward dissolved heavy metal cations at alkaline pH [1–4 and many others]. Several researchers have confirmed the potential of various forms of precipitated iron oxyhydroxides (PIO) and iron oxides in removing dissolved heavy metals to a much lower level than achieved by precipitation [5–7]. Both electrostatic (i.e., ion exchange) and Lewis acid–base (i.e., metal–ligand) interactions are the underlying reason for heavy metals' high affinity toward PIO over

295

alkali- and alkali–earth metal cations. In a simplistic way, the ion exchange type sorption of dissolved heavy metals (Me^{2+}) on PIO may be
presented as follows:

$$2(\overline{\equiv SO^-})Na^+ + Me^{2+}(aq) \leftrightarrow \overline{(\equiv SO^-)_2 Me^{2+}} + 2Na^+(aq) \qquad (1)$$

The overbar denotes the solid phase and $\equiv SO^-$ denotes a deprotonated
surface site present in any of the iron oxyhydroxides. Figure 1 presents a
visual interpretation of the forces involved in the metals sorption process
where the oxygen atom (Lewis base) may donate a lone pair of electrons
to a transition metal ion (Lewis acid).

Understandably, PIO would be an attractive sorbent in fixed-bed
columns for metal removal processes because they are likely to be much
less expensive compared to polymeric chelating exchangers often used for
the same purpose [8,9]. In spite of their desirable attributes, however,
commercial use of PIO in fixed-bed column has been practically nonexistent for the following reasons:

The current process of PIO synthesis, typically giving rise to a composition of ferrihydrite, goethite, and hematite, although straightfor-

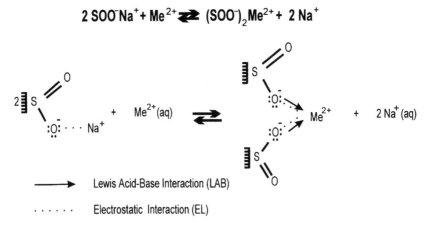

Figure 1   Schematic representation of heavy metal sorption onto deprotonated surface site.

ward, produces only very fine particles (microns), which are unusable in fixed beds because of excessive pressure drop in columns.
The uptake of dissolved heavy metals onto PIO is strongly dependent on pH and almost negligible at pH values below 5.0, as demonstrated by various researchers [10,11]. Consequently, PIO are ineffective for treating metal-contaminated solutions at pH values under 5.0 due to formidable competition from hydrogen ions, as depicted in the following equation:

$$\overline{2(\equiv SO^-)Na^+} + Me^{2+}(aq) + 2H^+(aq)$$
$$\leftrightarrow \overline{2(\equiv SOH)} + Me^{2+}(aq) + 2Na^+(aq) \tag{2}$$

At alkaline pH, on the contrary, most of the heavy metals are insoluble. Thus, use of PIO as heavy metal sorbent is limited to a polishing step after the precipitation/settling process.

In the recent past, it was demonstrated that sand, coated with precipitated PIO followed by heat treatment, enhanced iron oxyhydroxide microparticles, and flocculated iron oxyhydroxide microparticles, may be used effectively in packed-bed columns [12,13]. In both cases, very dilute acids at pH around 3.0 were used for regeneration. The foregoing morphologies (coated sand grains and enhanced microparticles) significantly improved the possibility of using PIO in fixed-bed columns, but they are still not suitable for treating heavy metal–contaminated wastewater at acidic pH (<5.0) and at relatively high concentrations of dissolved heavy metals, specifically mg/L versus µg/L. Also the iron oxyhydroxide–coated sand yielded very low metal-removal capacity even at alkaline pH because sand particles themselves did not possess any active metal sorption sites [12].

As mentioned, in the absence of competing hydrogen ions (i.e. at above-neutral pH) the sorption affinity of PIO for heavy metal ions increases significantly. Therefore, conceptually, if iron oxyhydroxide is modified in a way to ensure that the microenvironment around its sorption sites remain at alkaline pH, regardless of the bulk solution pH, very high metal-removal capacity can be achieved. Since the sorption process is kinetically much faster than precipitation, any precipitation of metal hydroxide within the pores of such a hybrid particle is quite unlikely while unoccupied sites are available [14,15].

A. Characterization of Hybrid Iron-Rich and Synthesized Iron-Rich Sorbents

The idea of a synthesized iron-rich sorbent (SISORB) evolved as a result of hybrid inorganic sorbent (HISORB) studies conducted by SenGupta, Ramesh, and Gao of Lehigh University from 1991 to 1995 [14,15,17]. The development of HISORB is a by-product from a process to reclaim zinc from electric arc furnace (EAF) dust. The EAF dust is a waste product produced as a result of recycling scrap steel using an EAF process. Steel is produced one of two ways, either from the raw components (limestone, iron ore, and coal) or from recycled scrap steel.

1. The Birth of HISORB and SISORB

The EAF method for the production of recycled ferrous scrap has resulted in the global production of more than one million tons of metal-laden dust annually [14]. In addition to ferrous-type metals present in the dust, it also contains significant quantities of zinc, lead, and cadmium volatilized during the steel-making process. In particular the zinc concentrations present in the EAF dust are very high due to the galvanizing process of steel used to reduce the corrosion process of raw steel.

Because the EAF dust does not pass the U.S. Environmental Protection Agency (USEPA) toxicity characteristics leaching procedure (TCLP) outlined in the Resource Conservation and Recovery Act (RCRA) of 1980, it is classified as a hazardous material and therefore cannot be directly landfilled. Due to this classification the EAF dust must be processed using an approved "best demonstrated available technology" (BDAT) (a method that ensures the dust will not leach toxic metals) before it is landfilled. The use of the Waelz kiln, a high temperature, thermal-processing technology, has been specifically recognized by the USEPA as a BDAT for stabilizing the EAF dust (Fig. 2). This method has proven not only to be a way to stabilize the EAF dust, but also to recover a substantial amount of zinc, cadmium, and lead.

The Waelzing process begins by collecting and conditioning the EAF dust to a moisture content of about 10%. In parallel, a flux mixture

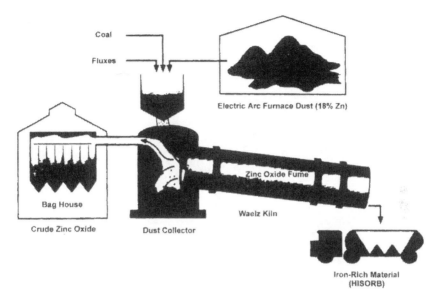

Figure 2 Waelzing process used to reclaim zinc.

of (CaO + MgO): SiO2 is maintained at a minimum ratio of 1.35:1 through the addition of dolomitic limestone and silicon dioxide. As shown in Fig. 2, the conditioned EAF dust is mixed with the fluxes as it is added to the blender. Approximately 1 ton of coal is added to every 4 tons of EAF dust that is charged to the kiln to keep the mixture from fusing into a solid mass as it passes through. The length of the Waelz kiln is about 150 feet long. The first 50 feet of the kiln serve as a drying and preheating zone with the remaining 100 feet carrying out the redox reactions.

The primary product from the Waelz kiln is an impure grade of zinc oxide containing 50 to 60% zinc. A co-product of the Waelzing of EAF dust is an iron-rich material which is virtually free of lead, cadmium, and zinc and exhibits no hazardous characteristic with respect to TCLP [16]. The iron-rich material, that is generated from the Waelzing process is what we refer to as HISORB. Table 1 is a list of the various elements that make up HISORB released in a Horsehead Resource Development, Inc. (HRD) 1989 report. A subsidiary of Zinc Corporation of America, HRD has been studying the possible use of the iron-rich waste product, HISORB, in the zinc reclamation process.

Table 1 Chemical Composition of
HISORB

Element	Weight, %
O	36.44 (% difference)
Fe	28.20
Ca	10.70
Si	8.40
Zn	4.20
Mn	3.60
Al	3.60
Mg	2.60
Cu	0.62
Na	0.55
Cr	0.31
K	0.31
Pb	0.18
Ti	0.17
Ni	0.12
Total	100.00

Work done by Gao and others [14,15,17] have shown HISORB to be an excellent sorbent, but its place on the market may never come due to its hazardous nature, as defined by law. Although the TCLP proves to be negative, as previously stated, the primary ingredient of HISORB, the EAF dust, has been identified as a hazardous waste under RCRA regulations. Because HISORB is a co-product of a hazardous waste material it is also considered to be a hazardous material under RCRA regulations. Because of the inevitable difficulties that would be encountered with legalization of the commercial use of HISORB, the idea of a sister product was envisioned, but without the impurities. The *synthesized iron-rich sorbent* created is referred to as SISORB.

2. HISORB

Hybrid inorganic sorbent materials were obtained in granular forms from Horsehead Resource Development Co. (Palmerton, PA). As described, a Waelzing treatment process is used wherein the particles that are formed pass through multiple solid phases during the process that takes place at approximately 1500°C in a reducing environment followed by a quenching process using water. The particles (porous slag) after quenching were subsequently ground to suitable sizes, between 100 to 500 microns, for use in fixed-bed columns. The particles in the final form are what our research group referred to as HISORB; HISORB is not a trade name.

Figures 3 a and b show scanning electron microscope (SEM) images of a typical HISORB particle used in this study. Note that by visual inspection of the images both crystalline and amorphous phases can be identified. X-ray diffraction of single particles identified the following crystalline forms: wustite (FeO), magnetite (Fe_3O_4), gypsum ($CaSO_4$), and akermanite ($Ca_2MgSi_2O_7$). A series of HISORB particles (not shown here) suggested that each particle is an aggregate of microparticles containing large macropores of sizes in the order of 10,000 Å [17]. The specific gravity of the particles is 3.5, while the average surface area is approximately 10 m^2/g.

(a)　　　　　　　　　　　　　　　(b)

Figure 3 Scanning electron microphotographs of a single HISORB particle with a magnification of (a) 100x and (b) 10,000x.

Of the major solid phases in HISORB, gypsum acts as a filler material. The oxides of iron (wustite and magnetite and other amorphous forms present) provide sites for ion exchange activity involving dissolved heavy metals. The final crystalline phase identified as akermanite, through X-ray diffraction, provides an elevated pH as it slowly releases hydroxyl ion in accordance with the following incongruent hydrolysis reaction:

$$\overline{Ca_2MgSi_2O_7}(solid) + 3H_2O \leftrightarrow 2Ca^{2+} + Mg^{2+} + 6OH^- + \overline{SiO_2}(solid) \tag{3}$$

3. SISORB

The synthesized iron-rich sorbent, SISORB (SISORB is not a trade name), is prepared using two active components:

1. Precipitated iron oxyhydroxides (PIO)
2. A mineral-like component similar to akermanite

In addition to the two active components a third component, calcium carbonate, is added to enhance the macroporosity of each SISORB particle. The two active components of SISORB are made in parallel, mixed together along with the third component, calcium carbonate, and fused at a temperature of about 2000°C. The resulting mass of incombustible material is termed a clinker. The clinkers that are produced are then crushed to the desired size, typically between 250 to 500 microns, and conditioned for use.

The completed SISORB particles look very much like the HISORB particles even though the specific gravity and surface area are somewhat different. The specific gravity of the SISORB particles is about 4.3, while the average surface area is approximately 1.175 m^2/g. It is interesting to note that although the surface area varies between the two particles on the order of one magnitude (10–1.175 m^2/g), the performance of the two inorganic sorbents proves to be very similar. Also similar to HISORB, each SISORB particle appears to be an aggregate of microparticles containing a series of macropores, as shown in Fig. 4a. X-ray diffraction analysis did not show the same akermanite peak that was identified in HISORB, but peaks for both wustite and magnetite were identified in the final fused product. Visual inspection of a SEM image shows a crystalline

(a)

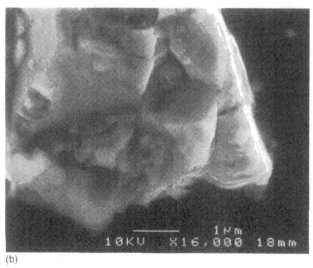

(b)

Figure 4   SEM imaging of SISORB.

structure similar to actual akermanite (Fig. 4b). Studies showed that the akermanitelike component behaved very much like actual akermanite.

4. Objectives of the Study

The primary objectivess were to

1. Synthesize an inorganic sorbent similar to HISORB
2. Present experimental evidence of the high metal sorption affinity of HISORB at an influent pH at and below 5.0
3. Show that SISORB could produce comparable results to HISORB
4. Compare the performance of the inorganic sorbents (HISORB and SISORB) with other sorbents, namely, polymeric chelating ion exchangers
5. Validate selective sorption as the predominate metal removal mechanism over precipitation
6. Explore in situ regenerability of the inorganic sorbents and their possible reuse

II. EXPERIMENTAL MATERIALS AND PROCEDURES

These studies include experimental results for four different sorbents: HISORB, SISORB, IRC-718, and Purolite C-106. HISORB was obtained from HRD, which uses electric arc furnace dust as the raw material in a specialty high-temperature (approximately 1500°C) process under reducing conditions to produce the waste product, HISORB. HISORB is essentially a by-product (i.e., waste product) of the zinc reclamation process. Prior to use in a column or batch titration, HISORB is rinsed with 100–200 bed volumes of distilled water in a column. A lengthy leaching test with deionized water, using a HISORB fixed-bed column, showed only traces of dissolved metals at concentrations less than 5 parts per billion. Various phases within HISORB were identified by using X-ray powder diffraction from a state-of-the-art automated Phillips System.

SISORB is synthesized using 60% precipitated iron oxyhydroxides, 15% calcium carbonate, and 25% synthesized akermanitelike component. The actual synthesis of SISORB will be discussed later in this chapter. SISORB was specifically engineered to remove cationic species, such as lead, cadmium, and other toxic heavy metals from contaminated water sources. Typically the material is crushed to a diameter between 425 to 250 microns (sieve sizes #40 and #60, respectively). It is then rinsed with 100–200 bed volumes of distilled water in a column.

The precipitated iron oxyhydroxides used in this study was prepared by titrating a 8.26 E-4 M ferric nitrate $[Fe(NO_3)_3]$ solution with a 1 N solution of NaOH. The PIO settled and aged between 2 to 4 hours. It was then separated using a centrifuge.

Once the PIO was separated it was then rinsed with deionized (DI) water and separated again using the centrifuge. The reason for rinsing was to wash out the NaOH still present in the matrix of the amorphous mass of PIO. It was rinsed with DI two more times. It was found that if the precipitate was rinsed more than three times the nature of the precipitate began to change, most likely due to the fact that as the pH of the amorphous mass begins to lower, some of the iron begins to go back into solution.

After the PIO has been properly rinsed it is then dried in an oven at $103°C \pm 5°C$. The dried PIO solids were then crushed using a mortar and pestle to a size of less than 75 μm.

A synthesized akermanitelike component (SALC) was prepared using specific amounts of calcium oxide (CaO), magnesium oxide (MgO), and silicon dioxide (SiO_2), at a ratio of $2:1:2$. The SALC is crushed to about 75 μm (Figs. 5a and b).

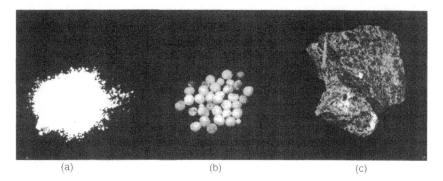

(a)   (b)   (c)

Figure 5   Various forms of akermanite: (a) crushed synthetic akermanite, (b) synthetic akermanite, (c) actual akermanite rock.

Naturally occurring akermanite rock was purchased from Ward's Natural Science Establishment, Inc. in order to compare with the SALC. Ward's, in turn, had obtained the rock from Dufresne Hill, which is located in Quebec, Canada, near Oka. The rock contains about 50% pure akermanite (Fig. 5C). The rock was crushed to about the same size as the PIO, 75 μm, and then used to compare the results regarding the natural akermanite verses the SALC.

IRC-718, a widely used chelating ion exchanger, was obtained from Rohm and Haas Co. in spherical bead form. They were conditioned following the standard procedure of cyclic exhaustion with 1 N hydrochloric acid and 1 N sodium hydroxide and finally converted to sodium forms for column runs. Chemical composition and other properties of IRC-718 are available in open literature [9,18].

Purolite C-106 is a weak base, organic, cation exchange material produced by Purolite Company, a division of Bro Tech Corporation (Philadelphia, PA). It is a spherical, macroporous bead with excellent resistance to osmotic shock. C-106 has a water retention of about 52–58% and a moist specific gravity of about 1.15 with a total volume capacity of 2.7 eq/L. It is conditioned following the standard procedure of cyclic exhaustion with 1 N hydrochloric acid and 1 N sodium hydroxide. C-106 is then converted to sodium form for use in all column runs unless specified otherwise.

A. Column Runs

All column tests were conducted using a glass chromatography column (250 mm effective length and 11 mm effective diameter). A constant flow pump was used to pass the solution through the column at flow rates varying from about 0.5 mL/min to 5 mL/min. To collect the effluent from the column runs an ISCO fraction collector was used.

To avoid channeling due to the wall effect, as proposed by LeVan and Vermeulen [19], a column diameter to exchanger bead diameter ratio of greater than 20 was maintained. Work done by SenGupta and Lim [20] showed that channeling due to the wall effect is practically absent under identical experimental conditions. The influent was pumped through the column in a downflow direction. pH was monitored at both influent and effluent by periodic sampling. Salient hydrodynamic parame-

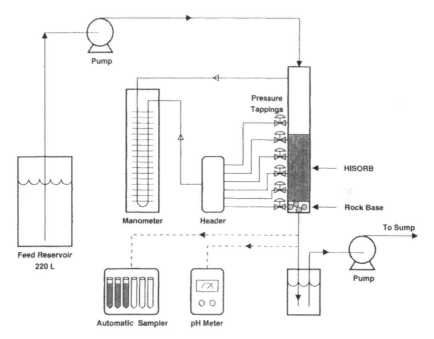

Figure 6 General schematic of HISORB packed-bed setup for the pressure drop study.

ters of various column runs, namely, empty bed contact time (EBCT) and superficial liquid velocity (SLV), are indicated in every figure.

Pressure drop studies and a number of supporting column runs were carried out in a larger 3-in. diameter column with 12 in. of HISORB (Nos. 16–48 ASTM mesh) supported on a 3-in. rock base as shown in Fig. 6. Pressure tapping from the top and intermittent bed depths were connected to a liquid manometer. The pressure drops for any two points within the bed during the column run were directly read from the manometer. At the end of every column run, the bed was backwashed simply by changing the valve arrangement. Connors [21] has provided detailed accounts of hydraulic behaviors of HISORB column runs elsewhere.

B. Acid Titration of HISORB

Acid titration was carried out by adding 1.0 g of HISORB in 200 mL Nalgene bottles containing aqueous solutions of different H^+ concentrations, but with constant ionic strength (0.1 M). The constant ionic strength was maintained by mixing 0.1 M NaNO3 and 0.1 M HNO_3 at varying proportions. The plastic bottles were equilibrated in a rotary tumbler and the aqueous phase pH was recorded after 1, 2, 3, and 7 days to develop the titration curve.

C. Chemical Analysis

Dissolved heavy metals, namely, zinc, nickel, cadmium, copper, and lead, were analyzed using an atomic absorption spectrophotometer (Perkin Elmer Model 2380) with flame attachment or graphite furnace accessory. Perkin Elmer's AANALYST 100 and SIMAA 6000 Spectrophotometers were used during the SISORB study. Sodium and calcium were also analyzed using an atomic absorption spectrophotometer in the flame mode. Sulfate and other anions were analyzed using Dionex Ion Chromatography systems Model 4500I and DX-120.

D. Surface Area and Pore Volume

Surface area and pore volume were measured using an Autopore III automated mercury porosimeter made by Micromeritics. The Autopore III has a pore range of 360 to 0.030 μm with a pressure range of 0.5 to 30,000 psi. Some samples were verified using an accelerated surface area and porosimetry analyzer, an ASAP 2405 using krypton as an adsorbate, also made by Micromeritics.

E. Particle Size Distribution

Two methods were used to measure particle size. One method used a number of standard ASTM E-11 specification sieves purchased from Fisher Scientific Company [#40 (425 m), #50 (297 m), #60 (250 m), #80 (117 m), #100 (150 m), and #140 (105 m)]. The material being measured was passed through the sieves at the beginning of the test and at the end of the test. The results of the distribution of sizes were compared and recorded graphically.

Verification of the results from the sieve analysis was done using a Beckman/Coulter LS 100Q particle counter with a patented binocular optical system. The enhanced laser diffraction system of the LS 100Q allows a particle size range from 0.40 to 1000 μm. The nature of the tests using the LS 100Q was the same as the sieve arrangement. The material was measured before each test and at the end of each test. The results of the distribution of sizes were compared and recorded graphically.

F. High-Temperature Furnace

To fuse the SISORB a high-temperature graphite furnace was used (Fig. 7). The actual furnace was a graphite furnace that was appropriated from a retired Perkin Elmer HGA-300 spectrometer. Although the amount of material made during each fusing was very small (less than 0.25 g) the furnace could be controlled very well and temperatures in excess of 2300°C could be reached.

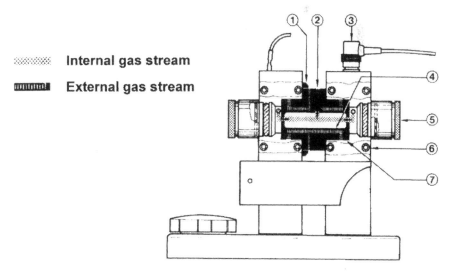

Internal gas stream

External gas stream

Figure 7   Graphite furnace: (1) graphite shield, (2) sample port, (3) temperature sensor, (4) graphite tube, (5) window assembly, (6) cooling pipe, (7) graphite cylinder.

III. RESULTS AND DISCUSSION

A. HISORB

1. Column Runs in the Presence of Competing Alkali Metal Cations (Na⁺)

A fixed-bed column containing HISORB material was fed with a solution of the following composition: $Na^+ = 250$ mg/L, $Cd^{2+} = Ni^{2+} = Pb^{2+} \cong 2.0$ mg/L, $Cu^{2+} = 0.1$ mg/L and pH 4.0. The feed solution was prepared using nitrate salts of metal ions. Figure 8 shows the effluent histories of all the dissolved heavy metals during the column run (concentrations are normalized with respect to influent concentrations). Almost complete removals of dissolved heavy metals were achieved up to 5000 bed volumes. Based on the order in which the breakthrough of heavy metals occurred, the selectivity sequence or binding strength of metal onto HISORB is as follows: cadmium < nickel ≪ copper < lead. The foregoing selectivity sequence is in agreement with the observations made by previous research-

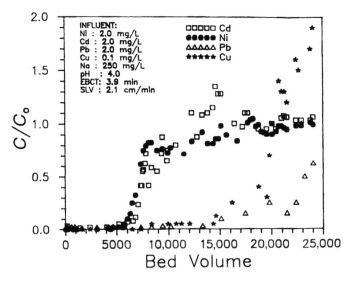

Figure 8 Heavy metals (II) effluent histories during a HISORB fixed-bed column run in the presence of competing alkali metal cations. C, effluent concentrations; C_o, influent concentrations; EBCT, empty bed contact time; SLV, superficial liquid velocity.

Table 2   Distribution Coefficients of
Individual Heavy Metals

Heavy metal (Me)	Distribution coefficient, λ_{Me} $\dfrac{\text{meq Me/g HISORB}}{\text{meq Me/mL solution}}$
Cd	3,500
Ni	4,400
Cu	9,050
Pb	12,100

ers using iron oxyhydroxides [11]. Also note that the three less preferred metal ions, namely, cadmium, nickel, and copper, underwent chromatographic elution after breakthrough (i.e., effluent concentrations were greater than the influent concentrations).

The distribution coefficients, λ_{Me}, for individual heavy metal ions, as computed from the dynamic column run, are provided in Table 2.

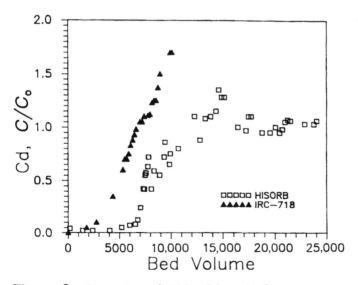

Figure 9   Comparison of Cd breakthroughs for two separate column runs using HISORB and IRC-718 under otherwise identical conditions. (From Ref. 15, courtesy of IWA.)

Very high distribution coefficients of all the dissolved heavy metals confirm their high affinities toward HISORB in the presence of much higher concentrations of competing sodium ions.

Figure 9 shows the comparison of cadmium breakthroughs for two separate column runs using HISORB and IRC-718. The influent composition and hydrodynamic conditions were identical for both column runs, as shown in Figure 8. Note that the cadmium removal capacity of HISORB is comparable to that of the chelating exchanger (IRC-718), and nickel underwent a lesser degree of chromatographic elution than copper, suggesting some possible precipitation within the column; however, cadmium and nickel were practically nondetectable up to 4000 bed volumes.

2. pH as a Surrogate Parameter for Metals Breakthrough

A column run was also carried out to evaluate HISORB's effectiveness in removing relatively higher concentrations of heavy metals (10 mg/L Zn) in the presence of competing calcium and sodium ions. Figure 10

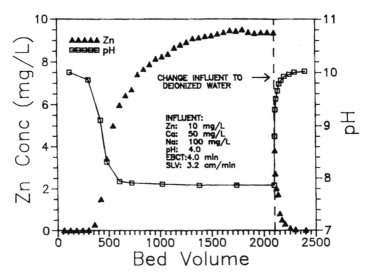

Figure 10 Zinc(II) effluent history and corresponding exit pH during a HISORB column run with influent replaced by deionized water at 2100 bed volumes. (From Ref. 15, courtesy of IWA.)

shows the effluent history of zinc and the corresponding exit pH during the column run. Note that soon after Zn(II) started breaking through from the column, the pH dropped significantly. This observation, that heavy metal breakthrough from the column accompanied a drop in exit pH, was observed for all other column runs where dissolved heavy metal concentration was 3.0 mg/L or greater. In order to confirm that such a column behavior (i.e., pH as a surrogate indicator of metals breakthrough) does not result from any other source/artifact, the column run in Fig. 10 was terminated after 2100 bed volumes, and the feed solution was replaced by deionized water. Note that the effluent pH quickly rose to 9.8, while the effluent zinc concentration dropped to less than 200 parts per billion.

The observation in Fig. 10 that zinc breakthrough from the column accompanies a significant drop in pH is amenable to explanation by considering sorption as the primary metals removal mechanism. When the dissolved zinc is being completely removed in the column before breakthrough, the effluent pH is almost equal to the pH obtained by hydrolysis of akermanite. However, when the sorption sites in the column tend to get saturated with zinc resulting in breakthrough, zinc forms labile mononuclear hydroxy complexes:

$$Zn^{2+}(aq) + nOH^-(aq) \leftrightarrow [Zn(OH)_n]^{2-n}(aq) \tag{4}$$

As a result, free OH^- concentration in the aqueous phase drops, thus reducing the effluent pH. Therefore, it can be summarized that as the zinc concentration rises in the effluent, zinc–hydroxy complexes are formed and this in turn lowers pH. When the feed was replaced by deionized water at 2100 bed volumes (Fig. 10) the pH rose quickly to 9.8 because there was no dissipation of hydroxyl ion in the absence of dissolved zinc. A model based on the stability constants of various zinc–hydroxy complexes can well predict the effluent pH at different zinc breakthrough concentrations (as shown by the solid lines in Fig. 13A–D).

3. Effects of Alkali- and Alkali–Earth Metal Cations and pH

Figure 11 shows effluent histories of zinc during the three separate HISORB column runs where the effects of competing ions were compared. In addition to maintaining constant hydrodynamic conditions in

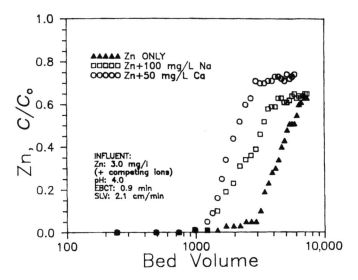

Figure 11 Zinc effluent histories during HISORB column runs in the presence of different competing ions. (From Ref. 15, courtesy of IWA.)

the three column runs, zinc was maintained at an influent concentration of 3.0 mg/L at a pH of 4.0. In the first column run no competing ion was present; in the second column run 100 mg/L Na^+ was present; and the third column run had 50 mg/L Ca^{2+} present as a competing ion. Note that the zinc removal capacity of the HISORB column decreased in the presence of the competing ions (i.e., zinc effluent history curves shifted to the left) and that divalent calcium ion, as expected, showed stronger competing effects compared to monovalent sodium ions.

In order to quantify the relative affinities of Zn(II), Ca(II), and Na(I) toward HISORB, separation factors were computed. The zinc/sodium separation factor for HISORB is dimensionless and defined as follows

$$\alpha_{Zn/Na} \equiv \frac{q_{Zn}C_{Na}}{q_{Na}C_{Zn}} \qquad (5)$$

q_{Na} and q_{Zn} are, respectively, zinc and sodium uptakes by HISORB in meq/g and C_{Na} and C_{Zn} are aqueous phase sodium and zinc concentra-

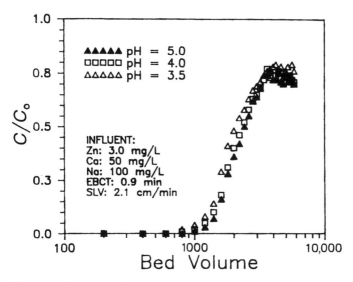

Figure 12   Zinc effluent histories during HISORB column for three different influent pHs under otherwise identical conditions. (From Ref. 15, courtesy of IWA.)

tions in the influent in meq/L. From the column run data, $q_{Na} = 0.0615$ meq/g, $q_{Zn} = 0.218$ meq/g, $C_{Na} = 4.35$ meq/L, and $C_{Zn} = 0.092$ meq/L. The zinc/sodium separation factor, $\alpha_{Zn/Na}$, was calculated to be 167. Similarly, Zn/Ca separation factor, $\alpha_{Zn/Ca}$, for the third column run was found to be 45.3. Such high separation factor values clearly demonstrate HISORB's effectiveness to remove dissolved heavy metals over alkali- and alkali–earth metals.

Figure 12 shows effluent histories of zinc for three column runs under identical conditions but with varying influent pH values (3.5, 4.0, and 5.0). Note that the difference in influent pH had no significant effect on zinc removal by the HISORB column.

4.   Multiple Cycles and In Situ Regenerations

In order for an inorganic sorbent to be viable for fixed-bed processes, it should be amenable to in situ chemical regeneration so that it can be used for a multiple number of cycles. After some trial and error, ammonia and another nonionized nitrogen-based ligand, ethylenediamine, were chosen

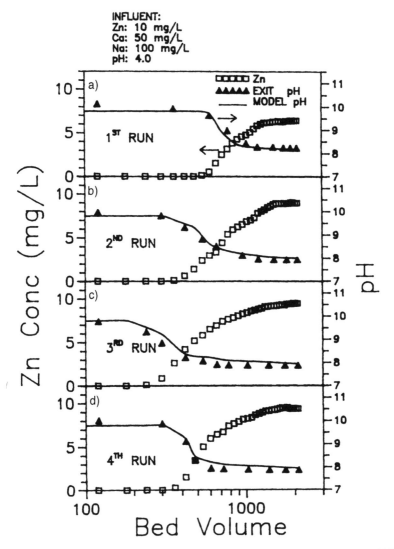

Figure 13 Zinc effluent histories and exit pH for four successive HISORB column runs. (Solid lines indicate model predictions of exit pH.)

as the regenerants. Figure 13a–d shows the effluent histories of zinc and exit pH for four successive column runs. Note that the zinc removal (i.e., the number of bed volumes of water treated before zinc breakthrough) remained fairly consistent for four successive column runs. Also, in accordance with the observation made in Fig. 10, note that the zinc breakthrough and drop in pH at the exit of the column occurred simultaneously.

Figure 14a–d shows the regeneration curves with ethylenediamine or ammonia after every column run. Weakly basic ammonia and ethylenediamine were chosen as regenerants because they do not chemically interfere with the akermanite phase in HISORB. Acid regenerants (both mineral and organic) are inappropriate because of their adverse effects on the buffer capacity of the accompanying akermanite phase.

5. Precipitation Within the Column

Since significant precipitation within the column may lead to operational problems due to increased pressure drop and channeling, experiments were carried out and observations made to ascertain such a possibility. The four consecutive column runs shown in Fig. 13A–D did not present any problem and precipitation was not observed visually within the column prior to zinc breakthrough.

In a separate test, a small amount of HISORB material (1.0 g) was loaded quickly with a high concentration (1000 mg/L) of zinc sulfate solution. Subsequently, the HISORB material was taken out from the column and stirred in a beaker to identify any loosely adhering zinc hydroxide precipitates. No such precipitate was observed in the liquid phase and the decanted liquid was very clear.

In order to confirm whether any precipitation within the bed would irreversibly affect the fixed-bed column, performance tests were conducted. A 12-in. deep HISORB bed in a 3-in. diameter column (see Fig. 6) was deliberately fed with a fairly high concentration of zinc (50.0 mg/L) and a differential manometer monitored the pressure drops between and the 4-in. and 7-in. bed depths. Figure 15a shows the zinc effluent history and the exit pH, while Figure 15b shows the pressure drop between the bed depths of 4 and 7 in. Note that the pressure drop increased gradually in the beginning and then took a sharp jump when the sorption capacity of the column was exhausted and the exit zinc concentration had

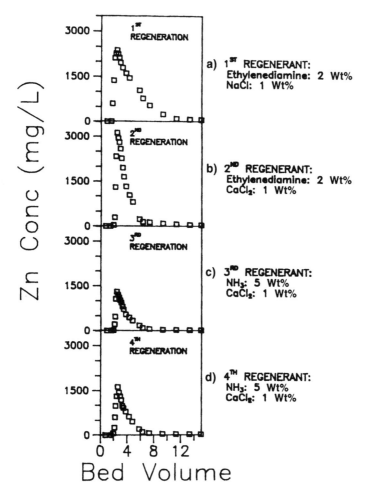

Figure 14 Zinc concentrations profiles during regeneration with ethylene-diamine or ammonia after every column run in Fig. 13.

risen sharply. This observation suggests zinc precipitation within the bed for such a high influent concentration.

At this point the column was backwashed using tap water (drinking quality) with approximately 20% expansion of the bed for 15 min. Subsequently, the column was fed with tap water for almost 300 bed volumes at the same liquid phase velocity (2.0 m/h). Note that the pressure drop

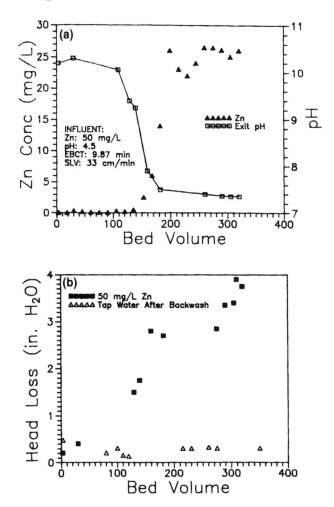

Figure 15 HISORB column behavior for an influent containing a relatively high concentration of zinc (50 mg/L): (a) zinc effluent history versus pH and (b) pressure drop between 4 and 7 in. of bed depth from the top. (From Ref. 15, courtesy of IWA.)

(Fig. 15b) was reduced markedly across the same bed depth (i.e., between 4.0 and 7.0 in.). The foregoing observation suggests that any metal hydroxide precipitation, if formed within the bed during the exhaustion cycle, is amenable to removal by conventional backwash and does not pose any irreversible fouling problem from an operational viewpoint.

6. Dissolution Studies with Akermanite

Dissolution studies were carried out with naturally occurring akermanite to confirm its ability to maintain alkaline pH in the aqueous phase through the incongruent hydrolysis reaction. About 0.1 g of powdered akermanite rock was added to 110 mL of distilled water at pH $\cong$ 5.4 and the solution stirred for 4 h. The pH was recorded with the aid of a pH meter. Subsequently, the slurry was filtered and the recovered akermanite was added to another 110 mL of fresh distilled water and stirred. Altogether three such dissolution cycles were carried out and the pH recorded. Figure 16 shows pH versus time for the three cycles. Note that a pH of 9.5 was attained very quickly even for the third cycle. Results of the dissolution studies suggest that, although sparingly soluble, akermanite may easily maintain alkaline pH through release of hydroxyl anions.

Similar dissolution studies were also carried out for virgin HISORB particles for three successive cycles under identical conditions. The change

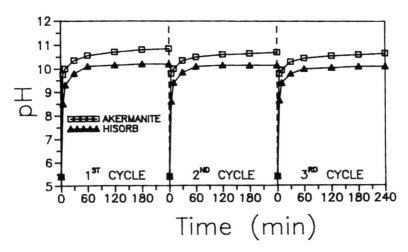

Figure 16 Comparison of akermanite and HISORB dissolution studies for three consecutive cycles.

Table 3 Equilibrium Ion Concentration for Three
Cycles of Akermanite Dissolution

	Mg^{2+}	Ca^{2+}	OH^-
First Cycle			
Concentration	2.88 mg/L	9.24 mg/L	0.0783 mMole
mMole	0.0130	0.0254	0.0783
Ratio	1	1.95	6.02
Second Cycle			
Concentration	1.90 mg/L	6.85 mg/L	0.0556 mMole
mMole	0.0086	0.0188	0.0556
Ratio	1	2.19	6.47
Third Cycle			
Concentration	1.72 mg/L	6.24 mg/L	0.0495 mMole
mMole	0.0078	0.0171	0.0495
Ratio	1	2.20	6.34

in pH versus time for HISORB is superimposed in Fig. 16; note that
the results for HISORB and natural akermanite are very similar. This
observation strongly suggests that akermanite in HISORB is the primary
ingredient that slowly releases hydroxyl ions over a long period of time
without being washed out from the fixed bed.

Based on the chemistry of other silicates [22], the stoichiometry of
akermanite hydrolysis is hypothesized as addressed in Eq. (3). Note that
for every 2 moles of calcium and 1 mole of magnesium, 6 moles of hydroxyl ions are introduced into the aqueous phase. Equilibrium calcium,
magnesium, and hydroxyl ion concentrations were determined during the
akermanite dissolution (three cycles) and are summarized in Table 3. Note
that the stoichiometry as predicted from Eq. (3) (i.e., $[Mg^{2+}]$: $[Ca^{2+}]$:
$[OH^-]$ = 1:2:6) is followed very closely by the experimental data validating the proposed dissolution.

7. Acid Titration of HISORB and the Interruption Test

Acid titration curves of HISORB in accordance with the procedure as
described earlier are presented in Fig. 17 for different periods of equilibration. The two following things are readily noted: first, HISORB has a
significantly high buffer or acid-neutralizing capacity (approximately 1.0–

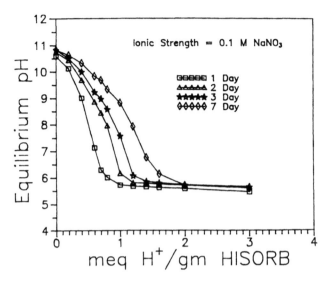

Figure 17 Acid titration curves of HISORB for different periods of equilibration.

2.0 meq H^+/g), and second, the buffer capacity increases with an increase in equilibration period. Also, the dissolved iron at the end of each titration curve was very low (less than 1.0 mg/L) indicating that dissolved iron did not contribute to the buffer capacity.

In order to gain an insight about the metal-removal mechanism and rate-limiting step of the metals removal process, a flow interruption test was carried out. A column run with a 3.0-mg/L influent zinc concentration maintained at a pH of 4.0 was stopped and restarted during breakthrough of the zinc. Figure 18 shows the effluent history of pH and zinc during a column run which was terminated at about 3600 bed volumes for a period of 24 h and then restarted. Note that following the 24-h interruption, pH in the effluent increased from approximately 8.0 to 10.0, while the zinc concentration dropped from 40% of the influent concentration to almost zero. However, the total zinc removal capacity of the bed did not increase because of the interruption and the effluent zinc concentration quickly rose to the level where it was before the interruption.

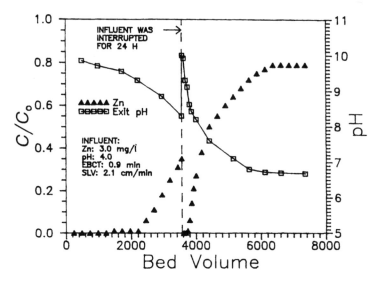

Figure 18 Zinc effluent history and exit pH during the column run with a 24 h interruption after 3600 bed volumes.

8. Role of Metals Sorption Versus Metals Precipitation

In the column run of Fig. 8, the dissolved copper concentration in the influent was 100 μg/L (parts per billion), which is below the solubility limit of total copper at all pH corresponding to the solubility product of the controlling copper hydroxide solid phase. Thus, any copper removal in the HISORB column cannot be attributed to precipitation. Figure 19 shows the exit pH and effluent history of copper during the column run. Also, superimposed on Fig. 19 is the theoretically calculated total copper concentration considering copper hydroxide precipitation within the column as the sole metal removal mechanism. By comparing the model with the actual copper removal it can be noted that far more copper is removed than can be achieved by precipitation; therefore, the removal cannot be a result of precipitation.

Figure 8 also shows that the four heavy metals (Cd, Ni, Cu, and Pb) do not break through concurrently. Cadmium, nickel, and copper breakthroughs undergo chromatographic elution (i.e., their concentra-

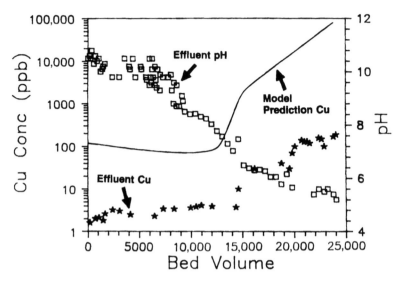

Figure 19 Comparison of effluent copper concentrations during a HISORB column run and the predictions from the equilibrium precipitation model. (From Ref. 15, courtesy of IWA.)

tions at the exit are greater than their influent concentrations). These observations are contrary to a precipitation mechanism, but can be well explained with multicomponent sorption or ion exchange theories with varying solute affinities toward the sorbent [23]. The competing sodium ion concentration in the influent was three orders of magnitude greater than the dissolved heavy metals, suggesting that selective sorption is the predominant removal mechanism for dissolved heavy metals.

For relatively high dissolved metal concentrations in the influent, and/or due to chromatographic elution effect, precipitations at elevated pH within the bed may occur. This is especially true for lengthy column runs. But precipitation is supplementary to the removal by sorption and the overall quality of treated water remains essentially unchanged because residual dissolved heavy metals are still sorbed selectively onto iron oxides. Also, heavy metals, if precipitated within the column, can be easily removed by conventional backwashing, as confirmed by the results in Fig. 15b.

The observation that the dissolved zinc concentration dropped to essentially zero during the column interruption test, as shown in Fig. 18, provides strong evidence that the hydrolysis of akermanite combined with intraparticle diffusion are likely to be the rate-limiting steps. This particular column behavior is an agreement with the interruption test results of other sorption processes, mainly ion exchange and ligand exchange, reported in the literature by Helfferich [24] and Zhu and SenGupta [25]. If precipitation were the primary metal-removal mechanism, effluent zinc concentration following the 24-h interruption would not undergo the same discontinuity, as exhibited in Fig. 18.

9. Overall Metal-Removal Mechanism

The simultaneous presence of akermanite and iron oxides in close proximity to each other in a single HISORB particle has a synergistic effect on the heavy metals sorption process. Figure 20 provides a schematic highlighting the fundamental differences in the metals sorption mechanism of the three sorbents, namely, the chelating exchanger (IRC-718), the iron oxyhydroxides, and the hybrid sorbent (HISORB) at a pH of 4.0. Note that every repeating iminodiacetate functional group of IRC-718 has two oxygen donor atoms and one nitrogen donor atom (tridentate) to satisfy the coordination requirement of the metal ion. As a result, the sorption affinity is relatively high and the hydrogen ion competition is less pronounced. Most of the chelating exchangers, therefore, show high metal-removal capacities even at pH values below 5.0 [26].

Iron oxides, on the other hand, have one oxygen donor atom per molecule and the zero point charge (ZPC) in the aqueous phase occurs at a pH around 7.2 [27]. This is the reason why hydrogen ion competition is quite fierce and the surface sites get overwhelmingly protonated at pH under 5.0, as shown in Fig. 20.

For HISORB, the underlying electrostatic and Lewis acid–base interactions remain very much the same as iron oxides, but the hydrolysis of neighboring akermanite eliminates hydrogen ion competition, thus enhancing metal-removal capacity markedly. The overall stoichiometry of the metals sorption process onto HISORB can be approximated by using the following equation involving both akermanite and surface sorption sites in iron oxides:

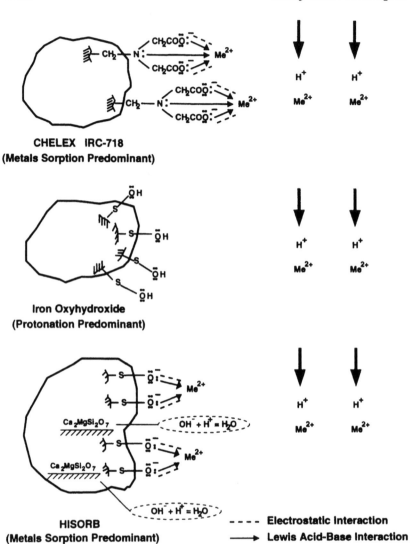

Figure 20 Illustration of the underlying mechanisms of heavy metals sorption onto three different sorbents at below neutral pH.

$$\overline{Ca_2MgSi_2O_7} + \overline{2(\equiv SO^-)Na^+} + 6H^+(aq) + Me^{2+}(aq)$$
$$\leftrightarrow 2Ca^{2+}(aq) + Mg^{2+}(aq) + 2Na^+(aq) \qquad (6)$$
$$+ 3H_2O + \overline{2SiO_2} + \overline{(\equiv SO^-)_2Me^{2+}}$$

Regeneration of metal-loaded HISORB with ammonia/amine in the presence of other electrolytes does not interfere with akermanite, but desorbs heavy metals through ion exchange accompanied by complexation:

$$\overline{(\equiv SO^-)_2Me^{2+}} + 2Na^+(aq) + nNH_3(aq)$$
$$\leftrightarrow \overline{2(\equiv SO^-)Na^+} + [Me(NH_3)_n]^{2+}(aq) \qquad (7)$$

B. SISORB

1. Synthesis Process

The two active components of SISORB are

1. Precipitated Iron Oxyhydroxides (PIO)
2. A mineral like component similar to Akermanite

A third component, calcium carbonate, is added to introduce macropores within each SISORB particle. The two active components of SISORB are made in parallel, mixed together along with the third component, calcium carbonate, and fused at temperature of about 2000°C for less than a minute. The resulting mass of incombustible material is termed a clinker, as defined by Webster [28]. The clinker is then crushed to the desired size, typically between 250 to 500 microns, and conditioned for use.

a. Precipitated Iron Oxide Precipitated iron oxyhydroxides (PIO) are actually a combination of various iron oxyhydroxides and oxides, therefore the general term PIO has been assigned to this component of SISORB. The following is a review of the primary iron oxide phases present at various times in PIO.

FERRIHYDRITE When ferric nitrate or ferric chloride is dissolved in water and precipitated as ferric hydroxide by raising the pH to about 11 using sodium hydroxide, the ferric hydroxide form is often short lived upon exposure to air [29,30]. There is a gradual transformation of iron hydroxide through rapid hydrolysis and nucleation into a predominantly amorphous state, commonly referred to as ferrihydrite. Ferrihydrite particles have been reported to be somewhat spherical, ranging in sizes from 1 to 10 nm [31–36]. Figure 21 shows a SEM image of what is presumed to be the amorphous state of ferrihydrite magnified 10,000 times. Each oval-like mass in Fig. 21 is made up of hundreds of the spherical ferrihydrite particles. By comparing the size of the spherical particles to the scale at the right base of the image, it can be determined that the size of each sphere is in the range of 10 nm or less.

Measurements of the density of ferrihydrite range from 2.2 to 4.0 g/cm^3, with an average at about 3.5 g/cm^3 [31,37–39]. The bulk structure of ferrihydrite is uncertain and its chemical composition is represented by the general stoichiometric formula $Fe_2O_3 \cdot nH_2O$ [40]. Thermogravimetric analyses of dried ferrihydrite samples have yielded n values between 1 and 3 [31,41,42] as have analyses made with X-ray photoelectron spectroscopy [42]. Towe and Bradley [43] have identified ferrihydrite as

Figure 21 Freshly prepared ferrihydrite (10,000x).

$Fe_5HO_8 \cdot 4H_2O$. In another study, by Chukhrov et al. [44], the chemical formula of ferrihydrite is postulated to be $Fe_6(O_4H_3)_3$. Schwertmann and Cornell [29] concluded the formula to be $5Fe_2O_3 \cdot 9H_2O$, while Russell [45] resolved the formula to be $Fe_2O_3 \pm 2FeOOH \cdot 2.6H_2O$. For the sake of simplicity this study will use the chemical formula of ferrihydrite as presented by Towe and Bradley, $Fe_5HO_8 \cdot 4H_2O$. From a mass balance perspective it is important to maintain an understanding of the stoichiometry. Equations (8) and (9) express the transformation from ferric nitrate to iron hydroxide to ferrihydrite:

$$Fe(NO_3)_3 + 3NaOH \leftrightarrow Fe(OH)_3 + 3NaNO_3 \qquad (8)$$

$$5Fe(OH)_3 \leftrightarrow Fe_5HO_8 \cdot 4H_2O + 3H_2O \qquad (9)$$

GOETHITE   Extended aging in an aqueous solution at 20 to 30°C gradually transforms ferrihydrite to a crystalline iron oxide, usually goethite (α-FeOOH). The rate of this transformation is faster in solutions of high Fe(III) content. At high pH values (>10) significant amounts (2 to 10%) of goethite appear in ferrihydrite samples after 12 to 15 days of aging [36,46]. Goethite is also formed through a process of thermal dehydration at temperatures of less than 70°C [29]. The sample in Fig. 22 was prepared from the same sample as shown in Fig. 21, but the pH was raised to about 11 and heated to a temperature of 60°C for a period of 24 h. There is a clear demarcation between the lighter color amorphous mass present, presumed to be ferrihydrite, and the newly formed short crystalline structures presumed to be goethite. The transformation can be visually detected as ferrihydrite changes from a dark brown color to an identifiable yellowish brown indicating goethite to be the prevalent from [29]. Equation [10] relates the stoichiometric relationship between ferrihydrite ($Fe_5HO_8 \cdot 4H_2O$) and a crystalline iron oxyhydroxide similar to goethite (α-FeOOH):

$$Fe_5HO_8 \cdot 4H_2O \leftrightarrow 5FeOOH + 2H_2O \qquad (10)$$

HEMATITE   There are a number of ways to prepare hematite. The most convenient and common methods, as outlined by Schwertmann [29], are

Figure 22 Ferrihydrite and goethite (10,000x).

1. Thermal dehydration, which converts a crystalline iron oxyhydroxide to hematite (Fe_2O_3) as expressed in

$$2FeOOH \rightarrow Fe_2O_3 + H_2O \tag{11}$$

2. Forced hydrolysis of acidic Fe^{3+} solutions,
3. Transformation of ferrihydrite in aqueous solution

The color of hematite is a reddish orange. Figure 23 is a sample that was prepared using thermal dehydration at 100°C. The majority of crystalline structures present in this image are much longer than in Fig. 22. The longer crystalline structures are presumed to be hematite, verified by the reddish orange color of the dried power.

Method for Producing PIO   Ferric nitrate nonahydrate [$Fe(NO_3)_3 \cdot 9H_2O$], purchased through Fisher/ACROS, was used to produce the PIO and then to synthesize the SISORB. Typically 100 g of the ferric nitrate

Figure 23   Hematite (10,000x).

nonahydrate was dissolved into 1 L of distilled water. Once dissolved, the pH was raised to 11 by adding a 1 N solution of sodium hydroxide, thereby causing the iron to precipitate out. The color of the solution would turn dark brown immediately, indicating the prevalent oxide to be ferrihydrite, as discussed. The freshly prepared ferrihydrite was aged for approximately 2 h in solution then separated from the supernatant. The separation of the supernatant from the precipitate was performed using a centrifuge. The precipitate was then placed in ceramic dishes and dried in an oven at 100°C for a period of 24 h. The color of the precipitate changed from an initial dark brown to a final color, which varies slightly from a yellowish orange to a reddish orange. The disparity in color indicates that the final dried precipitate is a variety of iron oxyhydroxides and iron oxides, therefore the term given to the dried precipitate is precipitated iron oxide. The PIO is then crushed to a size less then 75 μm, measured with the aid of a #200 sieve. Visual identification, with the use of a Color Plate [29], indicated that the most prevalent oxide and oxyhydroxide present to be hematite with a small amount of goethite, less than 5%.

b.   Synthesized Akermanite   As previously discussed, akermanite is a mineral made up primarily of calcium oxide (CaO), magnesium oxide (MgO), and silicon dioxide (SiO_2). Akermanite is capable of releasing excess hydroxyl ions over a long period of time as water is passed over it. The excess hydroxyl ions are in turn responsible for raising the pH. Through the dissolution process 6 moles of hydroxyl ions are released into solution for every 1 mole of akermanite dissolved, as shown in Eq. (12). From studies done by Levin et al. [47] the molar ratio of the oxides that make up akermanite has been determined to be 2:1:2, CaO, MgO, and SiO_2.

$$\overline{Ca_2MgSi_2O_7}(solid) + 3H_2O \leftrightarrow 2Ca^{2+} + Mg^{2+} \quad\quad (12)$$
$$+ 6OH^- + \overline{2SiO_2}(solid)$$

By mixing the oxides (CaO, MgO, and SiO_2) together in the appropriate molar concentrations and fusing the mixture in a graphite furnace at about 2000°C we have been able to synthesize a mineral very similar to actual akermanite. The akermanitelike component is referred to as synthesized akermanite; it is opaque, round, and solid, much like a small pebble when completed. The synthesized akermanite component is made in parallel with the PIO component. Once synthesized it is crushed to a size equal to the PIO, about 75 µm, and mixed together with both the PIO and calcium carbonate.

Figure 24 shows scanning electron microscope (SEM) images of both natural and synthesized akermanite. Figure 24a is an image, magnified 600 times, showing akermanite in its natural form. This particular akermanite sample was purchased from Ward's Natural Science Establishment, Inc., in the form of a rock. The particular akermanite rock was about 50% pure akermanite, and the remaining 50% various other inorganic compounds inherent to its origin in Quebec, Canada. The rock was crushed to about 75µm, the sample then prepared for SEM imaging, and the photograph taken. Figure 24b is a SEM image, magnified 8000 times which shows synthesized akermanite as previously described. Comparing the images, the shape and visible patterns appear to be very similar in nature.

(a)

(b)

Figure 24 Various forms of akermanite: (a) natural form (600x), (b) synthesized form (8000x).

c. Calcium Carbonate   Calcium carbonate is added in addition to the active components of SISORB in order to enhance the porosity of the sorbent once fused. At elevated temperature, calcium carbonate is converted to calcium oxide and gaseous carbon dioxide:

$$CaCO_3 \rightarrow CaO + CO_2 \text{ (gas)} \tag{13}$$

The theory follows that during the fusing of the three components—PIO, synthesized akermanite, and calcium carbonate—temperatures in the graphite furnace rise well above 1000°C. At this point the components begin to fuse together and the CO_2 gas evolves. The evolution of the gaseous CO_2 through the liquefied mass produces passages and pockets increasing the potential porosity, thereby enhancing the accessibility to surface functional groups.

2. The Final Product, SISORB

Once the two active components, PIO and akermanite, are prepared they are mixed together along with the calcium carbonate, now referred to as the SISORB mixture. The typical mixing percentages, by weight, are shown in Table 4. Once mixed thoroughly using a mortar and pestle, the SISORB mixture is passed through a #100 sieve to help disperse particles that may not be separating due to residual surface charges and moisture due to humidity. Next the mixture is placed in a 250 mL Nalgene polypropylene wide-mouth bottle, sealed, and set in a rotating batch receptacle for a 24-h period.

After the 24-h period the mixture is removed from the batch receptacle and it is ready for fusing. A graphite furnace was used to fuse the SISORB mixture. The graphite furnace used was reclaimed from a retired Perkin-Elmer Model HGA-300 atomic absorption spectrophotometer. The temperature is set at 2000°C with a ramp time of 4 s, hold time of 12 to 14 s, and a cool-down time of 10 s. The SISORB mixture is portioned out in small allocates, about 0.25 g or less, and placed in a graphite tube. Once the mixture is fused, what remains is incombustible; therefore the remaining mass is termed a clinker. A photographic study of the clinkers is shown in Fig. 25. The clinkers produced are anywhere from about 0.5 to 3.0 mm in diameter. Fusion at this high temperature produces a reducing environment. It is hypothesized that upon fusion of the SISORB mixture some of the iron(III) prevalent in the PIO is reduced to iron(II), which is found in the new iron oxide states formed, magnetite and wustite.

Table 4 SISORB Synthesis Scheme

	PIO	$CaCO_3$	CaO	MgO	SiO_2
Molar ratio	—	—	2	1	2
Weight %	60	15	25		
Function	Supplying selective metal sorption sites	(Porogen) Produces network of pores through evolution of CO_2	Synthesized akermanite—the local pH is raised as the synthesized akermanite dissolves.		

(a) Clinkers   (b) Size Comparison – Clinkers/Penny

(c) Crushed Clinkers Magnified 100x

Figure 25   Clinkers.

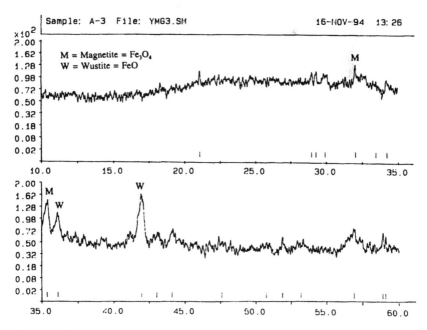

Sample: A-3 File: YMG3.SM   16-NOV-94 13:26

M = Magnetite = Fe_3O_4
W = Wustite = FeO

Figure 26   X-ray diffractogram of SISORB.

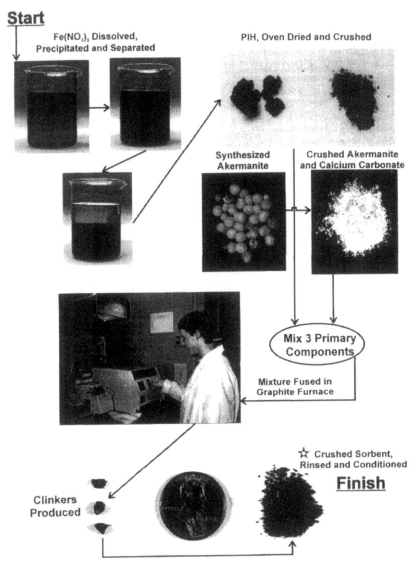

Figure 27 Overall synthesis process of SISORB.

Analysis using X-ray diffraction supports the presence of both magnetite and wustite in the final fused sorbent material, SISORB, as shown in Fig. 26. Equation (14) helps to put the final oxide transformation into perspective, assuming the transformation proceeds forward from hematite (Fe_2O_3):

$$4Fe_2O_3 \rightarrow 2Fe_3O_4 + 2FeO + O_2 \tag{14}$$

The clinkers are then crushed to diameters varying between 250 to 500 μm. Finally the SISORB particles are washed to remove any fine particles that could lead to head loss problems within a column and conditioned. Conditioning is done by passing 100 bed volumes of a 5% sodium chloride (NaCl) solution over the SISORB. The conditioning process insures that all available functional sites are loaded with sodium as the counterion, also referred to as the exchangeable ion. Figure 27 is a step-by-step overview of the entire synthesis process for SISORB [48].

C. Laboratory Tests

1. Comparing SISORB to HISORB

Figure 28 shows a comparison between SISORB performance and HISORB performance. Figure 28a represents data collected throughout three separate column runs using HISORB as the sorbent material. Each column run was conducted at different pH values (5.0, 4.0, and 3.0), while the concentration of metals remained the same. The metals present in the synthetic influent solution were zinc (Zn^{2+}), calcium (Ca^{2+}), and sodium (Na^+). Zinc, the target metal, was present at concentration of 3 mg/L, while calcium and sodium were present as competing ions at concentrations of 50 mg/L and 100 mg/L, respectively. The reason the column represented in Fig. 28a was run at three different pH values was to show how the akermanite component was able to control the local pH in such way that hydrogen ion competition, as previously discussed, was eliminated. With the elimination of hydrogen ion competition, the target metal, zinc, was removed effectively at all pH values.

The column run represented in Fig. 28b was conducted using SISORB as the sorbent material and all other conditions essentially replicating the column run done in Fig. 28a at a pH of 5.0. The superficial liquid velocity (SLV) and empty bed contact time (EBCT) in each separate col-

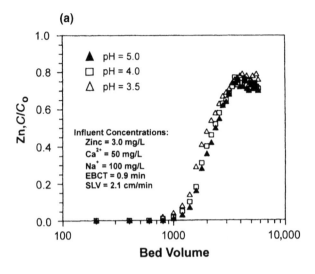

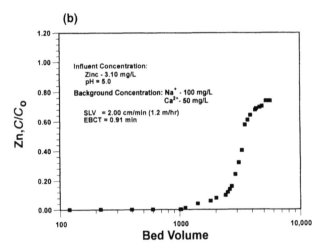

Figure 28 Comparison of zinc effluent histories: (a) HISORB, (b) SISORB.

umn run was maintained at about 1.2 m/h and 0.90 min, respectively. The target metal, zinc, and background metals, calcium and sodium, were also kept at about the same concentrations (3.0, 50, and 100 ppm, respectively) in each of the identified column experiments.

Experimental results presented by both graphs in Fig. 28 show that the behavior of SISORB is essentially identical to that of HISORB. Breakthrough for both the SISORB column and the HISORB column conducted at a pH of 5.0 shows to be at about 1000 bed volumes. Figure 29 offers a conceptual view of the sorption/dissolution process involved in metals uptake for SISORB. As an influent solution at a low pH value is passed over the sorbent material, SISORB, the local pH is immediately altered by the dissolution of the synthesized akermanite ($Ca_2MgSi_2O_7$). Because of the rise in pH caused by the release of hydroxyl ions (OH^-), competition from hydrogen ion is eliminated. The question, however,

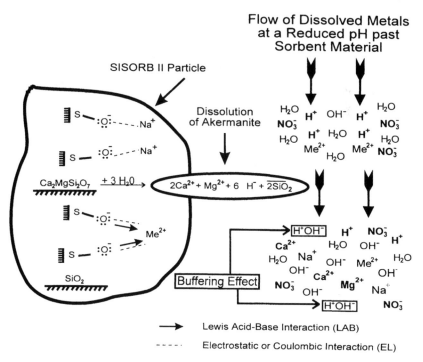

Figure 29 Conceptual view of SISORB sorption process.

arises whether precipitation plays any significant role in removing the target metal at this elevated pH. Figure 30 depicts such a possibility with sorption and precipitation as two parallel, simultaneously occurring metal-removal mechanisms. The following two sections will demonstrate the relative importance of precipitation versus selective sorption for the observed experimental results.

a. Evidence of Chromatographic Elution   Figure 31 shows evidence of chromatographic elution; sodium is eluted from the sorbent as zinc is taken up. Sodium starts out greater than 1, as represented by the C/C_o value in Fig. 31 (C = effluent concentration; C_o is the influent concentration). Therefore, this excess concentration of sodium would have to come from the sorbent, as the zinc is sorbed onto a site(s) presaturated with sodium. This would make the effluent sodium concentration higher than the influent concentration ($C/C_o > 1$). Once the sorbent capacity is reached with respect to the zinc, chromatographic elution of the sodium ceases. From the column run data reported in Fig. 31 this happens at about 1100 bed volumes. At this point the zinc continues to break through over a period of about 250 bed volumes and levels off at about 80% of the influent concentration.

b. Mononuclear Hydroxy Complexes   The column study represented in Fig. 31 has an influent pH of 5.0; the effluent pH starts above 9.5 and rapidly drops to about 8.0 as the zinc begins to break through. This observation (i.e., zinc breakthrough coinciding with a significant drop in pH) was also noted by Gao [14] for HISORB. Gao explained this phenomena with respect to the sorption mechanism as follows: ini-

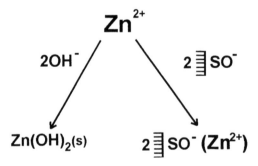

Figure 30   Precipitation or sorption?

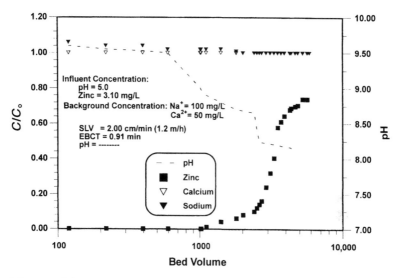

Figure 31   SISORB (presaturated with sodium ion) column run.

tially, when the zinc is completely removed by the sorbent material the effluent pH is about equal to the pH obtained by the hydrolysis of the synthesized akermanite. As the capacity of the SISORB is exhausted and zinc begins to break through, labile mononuclear hydroxy complexes are formed, as previously shown for HISORB in Eq. (4). As a result, the free OH^- concentration in the aqueous phase drops, thereby reducing the effluent pH.

c.   Selective Heavy Metal Removal   Ion exchange materials that have a functionality other than a strong acid group will typically show a range of selectivity toward various metals and nonmetals. The selective nature of SISORB is much like that of a cation exchange sorbent with a weak acid carboxylic functionality, where the oxygen donor atom plays a key role. The oxygen donor atom contributes both electrostatic and Lewis acid–base interaction with respect to transition metals. Figure 1 shows a representation of this interaction. Sodium, an alkali metal, exhibits only electrostatic interaction, but the transition metal, represented by Me^{2+}, exhibits both electrostatic and Lewis acid–base interaction. This combination of forces enhances the selective nature of the transition metal group (i.e., the heavy metal group), thus allowing the heavy metal to replace the sodium on the functional site. Each heavy metal has a different "selective strength" depending on the functional site, pH, and matrix of the ion exchange material. Therefore, it follows that when an ion exchange material that has a functionality such as that of SISORB or HISORB, a specific ion selective sequence should be seen as the heavy metal cations are separated from solution. This selective nature of the sorbent material helps to support the fact that Lewis acid–base interaction is the primary mechanism of separation. Figure 32 shows results from a column run using the sorbent material SISORB. In this particular column run the same background concentrations are used as in the column represented in Fig. 31, but instead of one target heavy metal, five different target heavy metals are present, each at an equal concentration of 2 mg/L. These results, along with other similar column runs, clearly indicate a reproducible, selective sequence with the sorbent material, SISORB. In this column run the sequence shows cadmium (Cd^{2+}) to be the least selective followed by zinc (Zn^{2+}). Nickel (Ni^{2+}) is more selective than zinc but less selective than copper (Cu^{2+}), and finally lead (Pb^{2+}) is the most preferred cation with

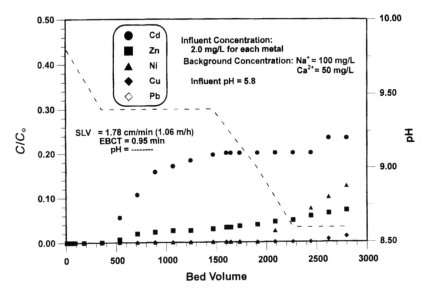

Figure 32 SISORB column runs with multiple number of heavy metals in the influent.

regard to SISORB. If the key mechanism were precipitation, this sequence in metal removal would not be seen. This information helps to strengthen the fact that the key mechanism to heavy metal removal using SISORB is selective sorption through complexation with SISORB's surface functional groups.

d. Visual Evidence The information presented so far helps to substantiate the position that the key mechanism to the removal of heavy metals with regard to SISORB is sorption rather than precipitation. Sometimes seeing is believing. To further reinforce this phenomenon, an additional "visual" study was conducted.

The test was done using two 1-L beakers, as shown in Fig. 33. Two liters of a zinc nitrate $[Zn(NO_3)_2]$ solution were prepared at an initial concentration of approximately 100 mg/L of Zn(II) using deionized water and were divided between the two beakers. Initial pH of the solution was about 4.0. Fifteen grams of SISORB were added to one beaker, the mix-

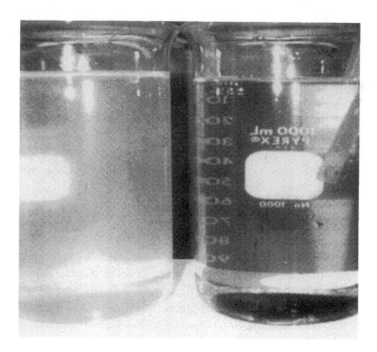

Figure 33 A visual comparison of sorption versus precipitation mechanism.

ture was continuously stirred over a period of 3 min, and, subsequently, SISORB particles were allowed to settle. The pH was measured at the end of the 3 min; it rose to about 8.0. The pH of the second beaker was then adjusted to about the same pH value of 8.0 using 1 N solution of sodium hydroxide (NaOH), but no SISORB was added. The solution in the right beaker, as seen in the photograph (Fig. 33), containing the settled SISORB particles remained clear and transparent. The beaker on the left, without any SISORB, turned cloudy as the pH was raised. The white flocs that formed in the second beaker were zinc hydroxide (i.e., zinc precipitate). The aqueous phase zinc concentration in the SISORB beaker, the beaker on the right, dropped to near-zero level. The SISORB beaker remained clear, precipitation was not visually detected, and therefore the zinc that was removed from the solution had to be sorbed onto SISORB particles. The beaker without the SISORB was filled with white flocs indicating zinc removal through precipitation. The contrast between the solutions of the two beakers (opacity versus transparency) at the same pH and with the same initial zinc concentration can be explained scientifically:

Due to its lower activation energy requirement, the kinetics of the sorption process is much faster than precipitation. Thus, in the presence of an excess amount of SISORB particles, metals removal from solution takes place primarily through sorption even at neutral to alkaline pH. Consequently, zinc hydroxide floc formation is minimal or absent altogether. On the contrary, in the absence of any SISORB particles in the beaker, flocs or precipitates will inevitably form as pH is raised. Solution will then turn cloudy. Rapid zinc removal by sorption is the primary reason why the solution in the right beaker did not form any flocs and thus remained clear and transparent even at elevated pH.

2. Benchmarking SISORB

To evaluate and benchmark the performance of SISORB pertaining to heavy metals removal, this particular study focused on a commercially available organic polymeric ion exchange resin, Purolite C-106. Figure 34 illustrates chemical composition of Purolite C-106, while Fig. 35 shows effluent histories of two separate column runs using SISORB and Purolite C-106. Concentrations of competing ions and influent pH were similar for both column runs reported in Fig. 35. Five different heavy metal ions were present in the influent at low concentrations. Note that

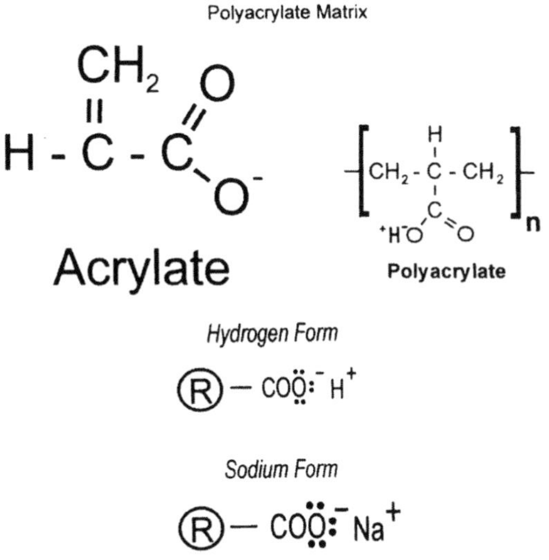

Polyacrylate Matrix

Acrylate

Polyacrylate

Hydrogen Form

$(R) - CO\ddot{O}:^- H^+$

Sodium Form

$(R) - CO\ddot{O}:^- Na^+$

Figure 34 Composition of weak acid cation exchange resin, Purolite C-106.

for both column runs, breakthroughs of heavy metals started around 1000 bed volumes. Other tests comparing commercially available ion exchange materials with HISORB produced similar results [17]. The experimental evidence in Fig. 35 confirms that SISORB can perform as well as, if not better than, commercially available sorbent materials.

Last but not least, many new materials have been identified or processed in the recent past for selective sorption of heavy metals. Materials such as seaweed [49], peat moss [50], minerals [51], metal oxides [52], sediment [53], kaolinite enhanced with leachate [54], humic acid [55], fulvic acids [56], and biomass [57] have all been found to have a significant ion exchange capacity for various toxic metals. As diverse as they are or may seem to be, they all have one thing in common. Weak-acid functional groups with oxygen donor atoms are the primary sorption sites for all these sorbents. That is why the underlying sorption mechanisms—competing effect of H^+ and relative selectivity of heavy metals—remain virtually the same for these sorbents. A linear free energy relationship (LFER) can, therefore, be a useful tool for predicting the relative selectivities of heavy metals for these sorbents (i.e., heavy metal selectivities are strongly correlated to aqueous phase stability constants of metals with

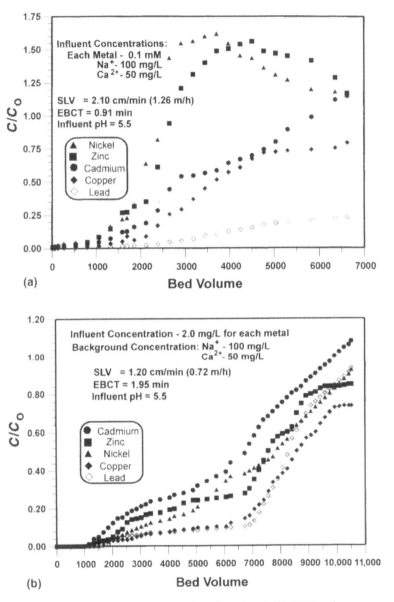

Figure 35 Comparison of Purolite C-106 and SISORB column runs: (a) Purolite C-106, (b) SISORB.

representative ligands containing carboxylate functional groups). Figure 36 shows a log–log plot of metal–acetate stability constants (K_{MAc}) versus metal/calcium exchange separation factors (K_{ex}) for three different sorbent materials, namely, ferrihydrite, Purolite C-106, and pea moss. Note that although absolute metal affinities vary from sorbent to sorbent, the affinity sequence (lead > copper > zinc) remains unchanged for all of them. From an application view point, removing lead or copper is thus more viable for all these sorbents than removing zinc.

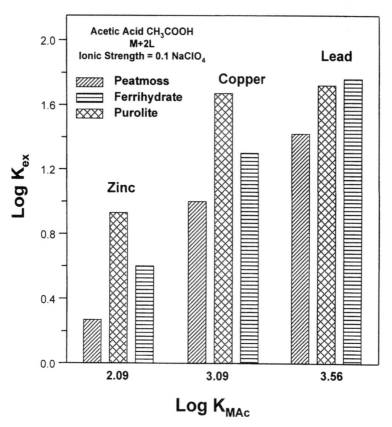

Figure 36 Heavy metal selectivity sequence for three sorbents containing surface functional groups with oxygen donor atoms.

IV. CONCLUSION

In this study, we undertook an extensive laboratory investigation to characterize, synthesize, and evaluate a new class of inorganic heavy metal sorbent. The primary findings resulting from this study can be summarized as follows:

HISORB and SISORB can effectively remove low concentrations (μg/L to mg/L) of dissolved heavy metals in fixed-bed processes with an influent pH as low as 3.5 from the background of much higher concentrations of competing calcium and sodium ions. Heavy metal loaded HISORB and SISORB particles may also be regenerated in situ using ammonia/amine solutions and thus can be reused for multiple numbers of cycles.

Comparative column run studies clearly indicate that both HISORB and SISORB can be effective as widely used as polymeric chelating ion exchangers [IRC-718 and DP-1 from Rohm and Haas Co. (Philadelphia, PA) and C-106 from Purolite Company, a division of Bro Tech Corporation (Philadelphia, PA)] in removing dissolved heavy metals.

Every single particle of HISORB/SISORB essentially contains iron oxide and akermanite. While the akermanite component helps eliminate the competition from hydrogen ions through slow release of hydroxyl ion, neighboring sorption sites on the various iron oxides quickly remove dissolved heavy metals. This synergistic relationship between akermanite and iron oxides make HISORB/SISORB effective inorganic heavy metal sorbents, even at acidic pH values.

For relatively high concentrations of dissolved heavy metals (around 50 mg/L) in the influent or due to chromatographic effect, removal by precipitation will occur in conjunction with sorption. The precipitate, however, can be easily removed by conventional backwashing.

Results of column interruption tests and the acid titration of HISORB particles suggested that hydrolysis of akermanite giving rise to hydroxyl ion is the most probable rate-limiting step in a fixed bed. Also, the effluent pH can be used as a surrogate indicator to monitor the breakthrough of heavy metals from the column for both HISORB and SISORB.

Commonalties do exist among the various sorbent materials outlined, namely seaweed, peat moss, minerals, metal oxides, sediment, kaolinite enhanced with leachate, humic acid, fulvic acids, and biomass containing negatively charged surface functional groups with oxygen-donor atoms.

According to the information in the open literature, HISORB and SISORB are the first inorganic iron-rich heavy metal sorbents that can be viably used in a fixed bed over a wide range of pH, as compared to others reported earlier [12,13]. At alkaline pH, heavy metals are insoluble while at acidic pH, iron oxides do not possess any metal-removal capacity due to protonation. As a result, use of iron oxides as a fixed-bed sorbent has, to date, been limited only to a polishing step after precipitation. The hybrid sorbent, HISORB, and the synthesized sorbent, SISORB, both provide new application opportunities to treat contaminated waters at acidic pH with relatively high concentration of dissolved heavy metals. In addition, the information presented on selectivity helps to develop a better understanding of the selective nature of sorbent materials with oxygen-donor atoms.

ACRONYMS

BDAT—best demonstrated available technology
BV—bed volumes
CPM—counts per minute
DI—deionized
EAF—electric arc furnace
EBCT—empty bed contact time
HISORB—hybrid inorganic sorbent
HRD—Horsehead Resource Development, Inc.
IMD—iminodiacetate
LAB—Lewis Acid–Base
LS—Liquid Scintillation
PHSAB—Principle of Hard and Soft Lewis Acids and Bases
PIO—precipitated iron oxyhydroxides
POTW—publicly owned treatment works
RCRA—Resource Conservation and Recovery Act
SALC—synthesized akermanitelike component

SEM—scanning electron microscope
SISORB—synthesized iron-rich sorbent
SLV—superficial liquid velocity
TCLP—toxicity characteristics leaching procedure
USEPA—U.S. Environmental Protection Agency
ZPC—zero point charge

ACKNOWLEDGMENT

The authors are thankful to Yi-min Gao and Anand Ramesh for their earlier research works on HISORB at Lehigh University. We also appreciate Ping Li's assistance in preparing several figures included in the chapter. Partial financial support for Arthur D. Kney's doctoral work received from the United States Environmental Protection Agency through Grant No. R 825422-01-0 is gratefully acknowledged.

REFERENCES

1. MM Benjamin. Environ Sci Technol 17:686–692, 1983.
2. PW Schindler. In: MA Anderson, AJ Rubin, eds. Adsorption of Inorganics at Solid–Liquid Interfaces. Ann Arbor, MI: Ann Arbor Science, 1981.
3. M Schultz, MM Benjamin, JF Ferguson. Environ Sci Technol 21:863–869, 1987.
4. DA Dzomback, FMM Morel. Surface Complexation Modeling. New York: Wiley-Interscience, 1990.
5. MM Benjamin, KF Hayes, JO Leckie. J Wat Pollut Control Fed 54:1472–1481, 1982.
6. M Edwards and MM Benjamin. J Wat Pollut Control Fed 61:481–490, 1989.
7. EA Forbes, AM Posner, JP Quirk. J Soil Sci 27:154–166, 1976.
8. D Clifford. In: FW Pontius, ed. Water Quality and Treatment. New York: McGraw-Hill, 1990, pp. 561–640.
9. WH Waitz. Amber-Hi-Lites 162, Philadelphia: Rohm & Haas, 1979.
10. CE Cowan, JM Zachara, CT Resch. Environ Sci Technol 25:437–446, 1991.
11. MM Benjamin, JO Leckie. J Coll Interf Sci 79:209–221, 1981.
12. M Edwards, MM Benjamin. J Wat Pollut Control Fed 61:1523–1533, 1989b.

13. TL Theis, R Iyer, SK Ellis. J Am Wat Wks Assoc 7:101–105, 1992.

14. Y Gao. Sorption enhancement of heavy metals on a modified iron-rich material. PhD dissertation, Lehigh University, Bethlehem, PA, 1995.

15. Y Gao, AK SenGupta, D Simpson. Wat Res 29:9, 2195–2205, 1995.

16. PL Kern, GT Mahler Jr. The Waelz process for recovering zinc and lead from steelmaking dusts. AIME Annual Meeting Technical Paper. No. A88-5, Phoenix, AZ, 1998.

17. A Ramesh. Selective sorption of heavy metals using an iron-rich waste by-product, Master Thesis, Lehigh University, Bethlehem, PA, 1992.

18. AK SenGupta, Y Zhu, D Hauze. Environ Sci Technol 25:481–488, 1991.

19. MD LeVan, T Vermeulen. Channeling and bed-diameter effects in fixed-bed adsorber performance. AIChE Symposium Series, No. 233, 80:34–43, 1984.

20. AK SenGupta, L Lim. AIChE J 34:2019–2029, 1988.

21. TM Connors. Hydraulic behaviors of an iron-rich heavy metal sorbent in a packed-bed column. MS Thesis, Civil Engineering Department, Lehigh University, Bethlehem, PA 1974.

22. RM Garrels, FT Mackenzie. Evolution of Sedimentary Rocks. New York: Norton, 1971.

23. FG Helfferich, G Klein. Multicomponent Chromatography: Theory of Interference. New York: Marcel Dekker, 1970.

24. FG Helfferich. Ion Exchange. Ann Arbor, MI: Xerox University Micro-films, 1961, pp. 250–322.

25. Y Zhu, AK SenGupta. Environ Sci Technol 26:1990–1998, 1992.

26. Y Zhu. Chelating polymers with nitrogen donor atoms, PhD dissertation, Lehigh University, Bethlehem, PA, 1992.

27. MM Benjamin. Effects of competing metals and complexing ligands on trace metal adsorption at the oxide/solution interface. PhD dissertation, Civil Engineering Department, Stanford University, Palo Alto, CA, 1978.

28. Random House Webster's College Dictionary. New York: Random House, 1991.

29. U Schwertmann, RM Cornell. Iron Oxides in the Laboratory. New York: VCH Publishers, Inc., 1991.

30. JD Bernal, DR Dasgupta, AL Mackay. The oxides and hydroxides of iron and their structural inter-relationships. Clay Min Bull 4:15–30, 1959.

31. AA van der Geissem. J Inorg Nucl Chem 28:2155–2159, 1966.

32. PV Anotins. Adsorption and coprecipitation studies of mercury on hydrous iron oxide. PhD dissertation, Stanford University, Stanford, CA, 1975.

33. PJ Murphy, AM. Posner, JP Quirk. Colloid Interface Sci 56:270–283, 1986.

34. E Tipping. Geochim Cosmochim Acta 45:191–199, 1981.
35. JHA Van der Woude, PL DeBruyn. Colloids Surface 8:55–78, 1983.
36. SA Crosby. Environ Sci Technol 17:709–713, 1983.
37. PJ Murphy, AM Posner, PP Quirk. Australian J Soil Res 13:189–201, 1975.
38. PJ Murphy, AM Posner, PP Quirk. J Colloid Interface Sci 52:229–238, 1975.
39. U Schwertmann, RM Tayler. Iron Oxides. In: JB Dixon, SM Weed, eds. Minerals in Soil Environments. Madison, WI: Soil Science Society of America, 1977, pp. 145–180.
40. FA Cotton, G Wilkinson. Advanced Inorganic Chemistry, 4th Ed., New York, John Wiley & Sons Inc., 1980.
41. DE Yetes. The Structure of the Oxide/Aqueous Electrolyte Interface. PhD dissertation, University of Melbourne, Melbourne, Australia, 1975.
42. DT Harvey, RW Linton. Anal Chem 53:1684–1688, 1981.
43. KM Towe, WF Bradley. J Colloid Interface Sci 24:384–392, 1967.
44. FV Chukhrov, BB Zvyagin, AI Gorshkov, LP Yermilova, VV Balashova. Ferrihydrite Izvest Akad Nauk SSSR Ser Geol 4:23–33, 1973.
45. JD Russell. Clay Min 14:190–214, 1979.
46. VL Snoeyink, D Jenkins. Water Chemistry. New York: John Wiley & Sons, 1980.
47. EM Levin, CR Robbins, M McMurdie. Phase Diagrams for Ceramists. Columbus, OH: The American Ceramic Society, 1964.
48. AD Kney. PhD dissertation. Department of Civil and Environmental Engineering, Lehigh University, Bethlehem, PA, 1999.
49. CJ Williams, RGJ Edyvean. Biotechnol Prog 13:424–428, 1997.
50. RH Crist, JR Martin, J Chonko. Environ Sci Technol 30:8:2456–2461, 1996.
51. X Chen. Env Sci Tech 31:3:624–631, 1997.
52. H Tamura, N Katayama, R Furuichi. Environ Sci Technol 30:4:1198–1204, 1996.
53. X Wen, Q Du, H Tang. Environ Sci Technol 32:7:870–875, 1998.
54. BK Schroth, G Sposito. Environ Sci Technol 32:10:1404–1408, 1998.
55. DG Kinniburgh, CJ Milne, MF Benedetti. Environ Sci Technol 30:5:1687–1698, 1996.
56. JA Leenheer, GK Brown, P MacCarthy, SE Cabaniss. Environ Sci Technol 32:16:2410–2416, 1998.
57. S Schiewer, B Volesky. Environ Sci Technol 30:10:2921–2927, 1996.

Index

353